The force of symmetry gives an elementary introduction to tł interplay between the three great themes of contemporary phy behaviour, relativity and symmetry. In clear, non-technical lan without oversimplification, it explores many fascinating aspe. physics, discussing the nature and interaction of force and matter.

Through the examination of relevant physical effects, and analogies from daily experience, the book presents in some detail the workings and implications of special relativity, quantum mechanics and symmetries. In so doing, the importance of these fields and their influence on the everyday world is highlighted. Towards the end of the book, its major themes are drawn together to describe the most successful physics theory in history, the 'standard model' of subatomic particles. The strange, counter-intuitive world of the very fast and the very small provides an excellent illustration of many of the topics discussed in earlier chapters.

The lively and non-technical approach of this book will make it suitable for first-year undergraduates in the physical sciences and mathematics, or those just about to embark on such courses, and for anyone with a general interest in these topics. It will also be a valuable accompaniment to more advanced texts on quantum mechanics and particle physics.

Vincent Icke is professor of cosmology at the University of Amsterdam and at Leiden University in the Netherlands. He studied theoretical physics and astronomy at Utrecht, and in 1972 obtained a doctorate at Leiden. He has held postdoctoral positions at the University of Sussex, Cambridge University, and the California Institute of Technology. After serving as a junior faculty member at the University of Minnesota, he came to Leiden in 1983. At present, his main interests are cosmology, the formation of structure in the Universe, astrophysical applications of chaos theory, and high energy hydrodynamics. Besides his academic pursuits, he takes an active interest in the popularization of science and in the arts. He is free-lance employee of VPRO Broadcasting and has participated in many productions on radio and television. In 1994 he was admitted as a student to the Gerrit Rietveld Academy for the Arts in Amsterdam.

THE FORCE OF SYMMETRY

THE FORCE OF SYMMETRY

VINCENT ICKE

Leiden University

'But if anybody says he can think about quantum problems without getting giddy, that only shows that he has not understood the first thing about them.'

Niels Bohr

CAMBRIDGE
UNIVERSITY PRESS

PUBLISHED BY THE PRESS SYNDICATE OF THE UNIVERSITY OF CAMBRIDGE
The Pitt Building, Trumpington Street, Cambridge CB2 1RP, United Kingdom

CAMBRIDGE UNIVERSITY PRESS
The Edinburgh Building, Cambridge CB2 2RU, United Kingdom
40 West 20th Street, New York, NY 10011-4211, USA
10 Stamford Road, Oakleigh, Melbourne 3166, Australia

First published 1995

Reprinted with corrections 1997

Printed in the United Kingdom at the University Press, Cambridge

Typeset in Monotype Times [TAG]

A catalogue record of this book is available from the British Library

Library of Congress Cataloguing in Publication data

Icke, Vincent.
The force of symmetry / Vincent Icke.
p. cm.
ISBN 0-521-40495-9. – ISBN 0-521-45591-X (pbk.)
1. Symmetry (Physics). I. Title.
QC174.17.S9I25   1994
539–dc20   94-26237   CIP

ISBN 0 521 45591 X paperback

Transferred to
Digital Reprinting 1999

Printed in the
United States of America

for Ieske
Without you, Earth is
an insignificant speck
in the Universe

Contents

Preface

The wings of a butterfly shimmer spectacularly in the sun. But when you pulverize these wings – please use only animals that have died of natural causes – the resulting powder is greyish-brown, not vividly coloured. That's quantum mechanics in action: the colour is not due to pigment, but it is due to the fact that light is made of particles that behave like waves.

Plants do not grow in the dark, even if you keep them warm. That's quantum mechanics in action: heat radiation, even if there is lots of it, cannot trigger the proper chemical reactions in leaves, because an individual particle of that radiation doesn't carry enough energy.

When you look out into the street through a window of an evening, you will see yourself reflected in the window, and yet the passers-by can see you, too. That's quantum mechanics in action: the same causes, apparently, do not always have the same effect.

Clearly, you do not need an expensive laboratory or a huge accelerator to see quantum effects. A healthy dose of curiosity will do very well; I think that it will in fact do much better, at a fraction of the price and at a billion times the enjoyment.

Physics is not difficult; it's just weird. Physics, contrary to the opinion of many journalists and parents of scientists, is not particularly hard to explain or to learn. To learn how to use relativity, you do not have to be Einstein; nor do you have to be Heisenberg to do quantum mechanics.

Physics is weird because intuition is false. To understand what an electron's world is like, you've got to be an electron, or jolly nearly. Intuition is forged in the hellish fires of the everyday world, which makes it so eminently useful in our daily struggle for survival. For anything else, it is hopeless. Our intuitive fear of heights would be ridiculous for an albatross; our intuitive appreciation of the flight of a ball is silly if we want to trace a quark. Intuition gives us plausible nonsense like astrology, homeopathy, or quantum-mechanics-turned-into-Zen. Intuition does not help us much in doing physics, be it quantum theory or classical mechanics (ever tried to understand the motions of a spinning top intuitively?)

Physics is powerful. That is a very good reason to write about it. Physics allows us to understand things: why the stars shine, why water is wet, why the sky is blue on Earth and orange on Mars. Physics teaches us to make things: from your contact lenses to the soles of your shoes, physics is at work. Physics is responsible for telephones, X-ray machines, gyroscopes, computers. Used wisely, like any power tool, treated with respect and understanding, physics benefits everyone.

Physics is beautiful. It makes me sad beyond words to know that so many people think of the physical sciences as barren, boring, bone-dry. Not so: when you lie outside in the grass on a clear dark night and look up at the stars, what you see is splendid. It is also physics. Understanding can lift you off the Earth, safer and faster and further than any rocket. The mind can travel among the stars, even enter them to see what causes those fires inside. To the beauty of seeing, we can add the beauty of understanding. And there is another level of beauty beyond that: the beauty of discovery, of creation, of *doing* physics. This beauty I love the most.

Physics is simplification, and so is explanation. Thus, insecure scientists need not fear that explaining physics to anyone is demeaning. The only thing that makes an explanation wrong (for experts or for novices alike) is if the reader has to unlearn something later. You will *not* have to unlearn anything when, having read this book, you go on to further study. This is the real stuff, even though there is a two-semester lecture series of heavy technical material behind every chapter. In fact, it is my intention that you could read this book right alongside a textbook on gauge field theory.

But no unlearning – just further reading. If I have to take anything back of what is in here, then either I blundered, or something really new has been discovered. Let us hope for the latter.

Physics may be easy to learn, but it is nearly impossible to do, to create. The weirdness of physics blocks the way. This prevents most of us from making new science, and it is what prevents most novices from understanding the old. Creating science is like creating poetry, of which Proust said: 'If it isn't easy, it is impossible'. To *do* physics, one has to be Bohr, or Curie, or Feynman. But people like me can write about it. And people like you can understand it (probably you're not like Bohr, but secretly I hope that you are).

But why this subject? I wanted to write about things that are well understood, by the standards of professional physics. That still leaves plenty of choice. I might have written about gravity, black holes, and all that, and someday maybe I will. I might even fit the words 'Einstein' or 'dinosaur' into the title. But for now, I am writing about matter and forces as they occur in the world of the very small, because, in a sense, the interaction of particles is the most physical of physics subjects. Black holes may be rare in our everyday world, but colourful butterflies, green plants, and shining stars are not. These common things do what they do because of the way their particles interact. Matter and forces perform a ballet; this book is a brief account of that amazing dance.

In the ten years it took me to write this book I have accumulated an enormous debt of gratitude to quite a few people. First there is Martinus Veltman with his profound, fascinating, chaotic, hilarious and inspiring lectures on fundamental physics. Second is Fran Verter who induced me to stop talking (no small feat) and to start writing (even harder). Third is an anonymous referee who caught a large variety of inaccuracies in the first draft and who advised CUP to go ahead with the book none the less. Then there's Donald Knuth with his awesome TEX typesetting language, the designers of the Macintosh computer and the folks who wrote the *Canvas* drawing software for it: I am amazed and humbled by people who can express clarity and beauty in bits and bytes. Next, Sheila Shepherd and Philip Meyler helped me along with their editorial advice and their inexhaustible patience. Dap Hartmann went through the last-but-one draft with a fine-toothed comb. In the final year the support of Frans Icke and Erica Ott was crucial. I hope that all of them will find that this book is a suitable payment of my debts.

Second printing

Many readers asked for the inclusion of more material on the mathematical properties of symmetry groups, and for more details on fundamental symmetries such as chirality. I am afraid, however, that too many potential readers have already been put off by what little mathematics there is in this book. Moreover, my knowledge of this part of the subject is far too shallow.

Numerous people kindly reported errors big and small in the first printing. I am grateful to David Broadhurst, Tony Hey, Jan Hilgevoord, Ray Lahr, Johan Lugtenburg and D.H. Rouvray for their contributions. I am especially indebted to Dolf de Vries for a no-nonsense lecture on basic chemistry and to Howard Chang for his very detailed criticisms. I also thank the latter for quoting Confucius: *If you believe everything in a book uncritically, then it would have been very much better if the book had never been written. The author is a human being and is likely to make mistakes and utter complete nonsense.*

Somewhat to my surprise, the most controversial statement in the book appears to be my contention *Physics is not difficult; it's just weird*. Maybe I should have omitted the qualification 'just', to emphasize my feeling than weirdness is a greater obstacle to understanding than anything else. But I stick to my guns. The fact that anything at all in the universe is comprehensible means either that we are very intelligent or that the basics of nature are very simple. Given that we are some sort of chimpanzee, carrying a mere kilogram of glop between our ears, I opt for the latter alternative.

Introduction

In the seventeenth century, great physicists and mathematicians like Descartes and Newton made the first real progress in understanding the interplay of matter and force, and in the description of the motions due to this. It must have been a marvellous epoch: for the first time, the workings of Nature on a grand range of scales were encompassed by theories that actually predicted things correctly. Since then, four forces have been discovered, and at present we are in the middle of another monumental advance: for the first time, there is a real prospect of a theory that explains *all* forces on the basis of one mechanism. We live in a marvellous epoch, and if you miss out on this revolution, you deprive yourself of a big piece of the action in the twentieth century. That is why I wrote this book: to instruct myself and to share with you the delight of this wonderfully intricate and powerful view of Nature.

This book is about experiences in the world that you and I live in, but the action takes place in an extremely special corner, which is not the one of our daily life. Thus, in reading what I have to say, your belief will be put to a severe test. What are you going to accept of all the extraordinary explanations that lie here before you? As stated in the splendid book by Abell and Singer,

'We accept that the light from stars in the night sky is thousands or even millions of years old, even though that concept contradicts our intuitions; we believe those same stars are unimaginably huge balls of glowing gases whose light is maintained by nuclear fusion [. . .] even though they look like tiny, pristine jewels. We believe that this book we are holding is solid, even though at the atomic level we know that it is almost 100 percent empty space'.

And then I come along, telling you that, on the subatomic level, this book *is* empty space, in a sense, and that everything we see can be reduced to the behaviour of a kind of stuff called 'vacuum'! All these things may seem bizarre, but to the best of today's knowledge they are true. You need not take my word (or anyone else's) for this; one of the fun things about physical facts is that you can verify them yourself if you want to spend the time and effort. This book is mostly about facts (and conjectures whenever we are dealing with frontline stuff); some will seem even more bizarre than those just quoted and are more difficult to verify.

The emphasis of this book is on the theory of matter and force; only incidentally will I mention specific observations, such as the mass or the decay time of a certain particle. I like the theoretical side of physics best, and theory has got rather short shrift in popular writing. By the way, it would be wonderful if someone were to write a parallel book that is *really* about accelerators, detectors, computers and all that. It ought to contain not just a few incidental pictures from bubble- and spark-chambers, larded with thumbnail sketches of personalities attending them, but should be packed from cover to cover with the all-stops-out high technology that is the glory of today's experimental science.

There are quite a few books that place more emphasis than I do on the experimental findings in particle physics. A reference list is given at the back of this volume. These books also contain more history than is presented here.

Because this book is about theory, it might be unsettling in several ways. First, the emphasis on theory will take us into somewhat uncertain territory, because this is a field in which there is an intense activity at present. Thus, you may encounter statements like 'This was thought to be absolutely impossible, but then someone realized that ....' Such is progress in theoretical physics. Second, you will probably find many things here that are radically new to you. I have tried to soften the impact of this, by introducing concepts first in the form of a sketch, for example in the discussion of the differences and similarities of 'matter' and 'force'. This sketch is filled in and detailed later, when more of the ingredients of the overall picture are available. Thus, you will sometimes want more information than is given a few chapters hence; if you find that you haven't got it by the end of the book, you're ready for reading the more technical work specified in the reading list. Third, unless there's a major bug in the typesetting software, this book does not contain the same stuff twice. Moreover, it follows a line of development and argument. So you will have to read it much more carefully than a coffee-table book. Maybe (horrors!) you'll even have to reread it.

I have tried to relate the current state of affairs in the theory of forces, but I did not write a book to convince you of anything: I have no philosophy to sell and I refuse to get bogged down in sterile speculation on what 'reality' means, and that sort of thing. For me, reality is what we have in common – you can do experiments and calculations just as well as anyone and possibly better. Thus, if you don't believe the physics, I demand that you study the original publications I used – there is a reference list – or, better yet, go and study physics yourself, and contribute to its advance.

Today's understanding of forces rests on three pillars: relativity, quantization and symmetry. These may seem formidable, but they aren't so bad; your worst obstacle is yourself, or rather the prejudices you acquired by being a certain size. On human scales of length, mass and speed, most effects that dominate the lives of subatomic particles are imperceptible, even though we'll see later that the very solidity of matter depends entirely on a quantum property. Our intuitions, and hence our philosophies, have a devastating bias against anything that lies outside the minuscule compass of gross everyday physics. Conquer the chauvinism due to slowness and bigness and you will be free to enter the spectacular world of the very fast and the very small, where relativity and quantization reign. Your reward will be a picture of forces that is, perhaps surprisingly, not too difficult to accept.

Our everyday experience of the world may delude us into thinking that 'classical' physics is obvious. But, in fact, almost nothing is self-evident. When we take a few steps down the street, we occupy an entirely different location in space and yet we do not change. When we turn a corner, we are not suddenly transformed into a giant penguin. This is so commonplace that it has ceased to amaze us, if indeed we ever stopped to notice. And yet we know of no compelling reason why these things should be so. In other words, we don't know why they are necessary in order to obtain a properly working Universe. Indeed, if you were to construct your own universe, you might well be able to arrange things differently; quite possibly, there are excellent reasons why space and time behave the way they do, but we haven't discovered them yet.

As the story unfolds, we will see that the most elementary question about the working of the Universe – namely *where is what when* – is answered in different ways, which reveal more and more about the way in which the Universe is put together. First, the three ingredients of the question ('where', 'when' and 'what') are considered in the form of classical mechanics, the theory of everyday life with which you are most familiar. In that case, each of these three items must be introduced separately: space for 'where', time for 'when' and particle for 'what'. In this picture, there is no connection or kinship whatever between these ingredients.

Second, the introduction of relativity forges a bond between 'where' and 'when'; thus, we learn to work with one ingredient, space-time, which takes over the role of what formerly was the duo 'where' and 'when'. Third, quantization allows us to give meaning to 'small' and 'large', which is necessary if we want to explain large and complex systems in terms of smaller and simpler ones. But we are compelled thereby to replace particles with `quanta` as entities of 'what'. Also, we discover that there is a basic uncertainty in the behaviour of a quantum, which brings us dangerously close to allowing chaos to reign the world. Fourth, the introduction of `symmetry` imposes order on the quantum chaos and enables us to merge all three ingredients into one: we find that quanta and space-time obey rules that are imposed by symmetries. Thus, the 'where', 'when' and 'what' are folded together into one, the `vacuum` (perhaps, in the most modern theories, this is a ten-dimensional kind of stuff). In this amazing picture, the properties of the vacuum encompass all the workings of the Universe.

I have suppressed almost all mathematics in the presentation, in part to accommodate my readers, in part to accommodate myself: although some facets of the relevant mathematics (especially those on symmetry) have great beauty, the nitty-gritty of relativistic quantum theory is often singularly repulsive (the cognoscenti who have seen, for example, the full expression of the electroweak Lagrangian, with its horrid tangle of typefaces and indices, may know what I mean). Even so, the stuff presented in this book is quite deep, but we mustn't confuse deep with difficult. Working out the quantitative details of the processes described here requires an enormous amount of difficult calculation, but the problem is often just technical. The depth of our subject lies not there, but in the insights of the physics. The basics of these I have tried to present; the obstacle to our understanding is not the technical complexity, but our hidebound intuition. What gets in our way is that the concepts are unusual, not that they are exceedingly difficult.

Even so, there are a few places in the text where I have used mathematical symbols. Most of the time, this is a matter of economy. It is very cumbersome to write 'Planck's constant' all the time when simply jotting down $\hbar$ will do. And I'm sure that you cannot be bothered to read that sort of verbose text. In some other places I have written a few elementary equations of the type $E = mc^2$. If such things *really* bother you, just skip them, and you will hardly be worse off. Or, preferably, just follow the text and re-read the occasional paragraph or two, just like all of us physicists do.

The price to be paid for leaving out the mathematics is that many things are said in the form of plausibility arguments or metaphors, and some things remain unsaid altogether: in those cases, precision is lost. This is sadly inevitable when using ordinary language. It is known that the great Niels Bohr struggled mightily with the problem of conveying physics in words, and for lesser mortals this is certainly impossible. To me, narrating physics is like casting shadows of a very complex object: one never gets the whole shape, but illuminating it from many different angles helps.

Plausibility arguments are all one can give at *any* level in physics, because the laws cannot be found by deduction. But there is always the danger of carrying a metaphor too far; in the end, we cannot escape the fact that to us, who are so very much heavier and bigger than particles, the world of the very small will forever be an alien place. I can well imagine that metaphoric writing annoys some readers, especially those with knowledge in the physical sciences. Yet I make no apologies: I'll be happy if this book induces them to acquire the necessary skills in maths and to pass on to the real stuff in the source texts. In fact, that would be a gratifying function of my song of praise for these wonders of Nature: just as a museum visit should induce you to practise art, I hope that my book will encourage you to study physics.

1

A matter of force

⭐

1.1 The law of inertia

The way the world works is mostly the way things move: *Where is what when?* is just about the most basic question one can ask about the Universe. Everyday experience gives us a rough-and-ready answer: the motion of matter is governed by forces. A puck may lie still on the ice until it is struck with a stick, after which it glides straight along until it hits something else. Without being struck, bumped, caught, or otherwise interfered with, it will follow its own path.

This description is horribly vague. On the ice, the puck moves with very nearly constant speed in a straight line. But the same object, struck in the same way, moves very differently on the pavement: almost as soon as the blow that sets it in motion is over, the puck lies still again. At the very least, then, it is unclear what an object's true path is: the smooth gliding along the ice, or the state of rest on the pavement, or what?

We cannot specify what we mean by 'force' until we have specified what ideal state of motion that force is supposed to perturb. Some four centuries ago, it was generally assumed that motion with constant speed along a circle is the ideal motion that can maintain itself indefinitely without external influence. This idea (although we now know it to be wrong) is not in itself absurd: we cannot deduce from first principles whether or not uniform circular motion is ideal in the above sense. Indeed, if we were to build our own universe, maybe we could arrange it that way.

But we *observe* that in the actual Universe a circular motion cannot maintain itself indefinitely, as is evident in the operation of a slingshot (Fig. 1.1). In another universe, constructed along the lines of the classical philosophers, maybe David's stone would have continued to buzz around in a circle, rather than fly off on a tangent to strike Goliath down.

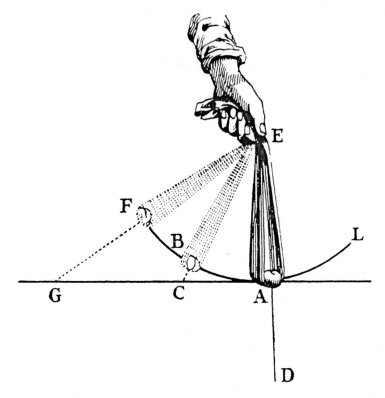

Fig. 1.1 Engraving illustrating the action of a slingshot, showing the law of inertia at work: the stone, when released from the constraining force of the sling, moves away in a straight line. (From Descartes's *Principes de la Philosophie*, part two, Art.39.)

A major advance towards today's theories of Nature was made by Descartes, who formulated what we now know as the Law of Inertia: *a piece of matter moves in a straight line with constant speed, unless a force acts on it.* This introduces force as something that causes the state of motion of an object to become different from the ideal constant-velocity motion. The pivotal question is then: what is that something?

Descartes himself thought that a force comes about through the immediate physical contact between objects. To anyone who limits the observation of Nature's workings to an occasional billiards game, this idea is self-evident. Descartes knew that the motions of the planets must be governed by some sort of force, even though they do not seem to be in contact with anything (pinned to crystalline spheres, or whatever). Planetary orbits are *curved*, so the law of inertia implies that there is a force which acts on them constantly.

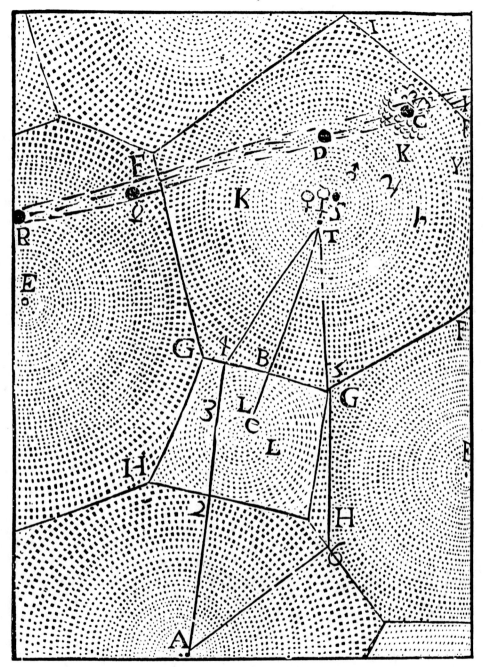

Fig. 1.2 Vortices in the 'subtle matter' or aether which, according to Descartes, was responsible for transmitting the forces that keep the universe together. (From *Le Monde, ou Traité de la Lumière*.)

Accordingly, Descartes assumed that all of space was filled, to the complete exclusion of emptiness, with a novel kind of substance, a subtle matter that transmitted the forces in the Solar System through direct contact. This matter was supposed carry the celestial bodies around in its swirling eddies (Fig. 1.2). In the second part of his *Principes de la Philosophie*, Article 16, Descartes wrote:

...from the sole fact that an object is extended in length, height and depth, we have reason to conclude that it is a substance, [and so] we must conclude the same about space which is supposedly empty: namely that, because it possesses [spatial] extent, it also has substance.

In other words: space has physical attributes, namely its three dimensions, so it must be regarded as real stuff. This powerful notion lay hidden for three hundred years, until it was rediscovered independently by Einstein.

One could make some objections of principle, for example that the constituents of Descartes's subtle matter must not have any internal structure, and hence must be infinitely small, wherefore – by Descartes's own definition – they have no spatial extent and consequently do not exist. Indeed, this problem had been spotted in antiquity, and gave rise to the hypothesis that the world is made of atoms: small particles that have no inner structure and therefore cannot be further divided, but that do have a finite extent. Of course, the atoms we now know do not fit that description at all, even though they have the same name. Later I will return in detail to the consideration of smallness.

Unfortunately for its inventor, the Cartesian hypothesis about the direct-contact origin of forces did not lead to calculable results. Descartes proposed that the planets are kept in their orbits by swirling, vortical motions in the subtle matter between them; but this assumption did not give a quantitative prescription for the behaviour of the force. Thus, nothing could be calculated; for example, Kepler's laws (which were known to describe planetary motion very well) could not be explained in terms of the motion of the subtle matter.

It was Newton who first realized that, for the description of planetary motions, it is not necessary to know what a force 'is', as long as one can give a precise description of what it does, i.e. formulate how it depends algebraically on physical quantities. In fact, it had already been pointed out by Hooke that Kepler's Third Law implies that the force between the Sun and a planet acts along the line connecting them and decreases in inverse proportion to the square of the distance between them. Newton extended this with the prescription that the force be proportional to the product of the masses of the attracting bodies. This was an important advance, because it established a symmetry between the objects involved. It isn't as if one object, for example the Sun, is the boss that does all the attracting; in the Newtonian description of a force, both objects attract *each other*, so that we can truly speak of an interaction. Moreover, in keeping with Descartes's hint that there is only one force in Nature, Newton presumed that the dominant force in the Solar System (to be known as gravitation) is universal and acts between all objects.

The success of this approach is well known, and it has been praised beyond measure. And yet the Newtonian idea of force had some uncomfortable features. It was conceived to be an instantaneous interaction: a mutual working, at exactly the same moment, between two spatially separated objects. The instantaneous nature of the action was not, at first, recognized as a problem; but the objections to the 'action at a distance', across supposedly empty space, were loud from the beginning. Still, it worked. Planetary motions could be calculated; Kepler's laws were explained in terms of the force of gravity. The Cartesian hypothesis of direct contact was completely eclipsed by the Newtonian action at a distance.

It was clear even in Newton's days that there must be more forces in Nature than gravity alone. Whereas Descartes's forces could appear in many guises – attractive or repulsive – depending on the detailed workings of the subtle matter, the gravitational force is always attractive, and hence cannot make stable objects: all things always fall down, so to speak. Thus, all many-particle systems in our Universe, when acted upon by gravity alone, must inevitably collapse, even though this might require a very long time. The apparent solidity and stability of matter is proof of the existence of other forces, so there was scope for an extension of the Newtonian system by finding those forces. Some were found, such as the magnetic and electrostatic forces; the experiments of Cavendish and Coulomb even showed that the electrostatic force can be described by exactly the same mathematical form as the force of gravity. But nobody questioned the underlying concept of instantaneous action at a distance any more.

1.2 The speed of light

A dramatic and fundamental step forward was made by Maxwell. This advance was wholly within the Newtonian world view but, interestingly, was also one of the first nails in the coffin of that view. Maxwell showed that the forces of electricity and magnetism are not two totally different ones, but instead are two aspects of one force (albeit a more complex one), the

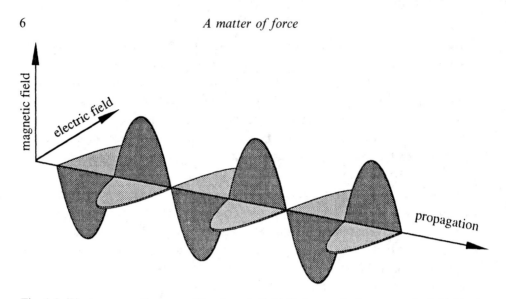

Fig. 1.3 Electromagnetic wave. The electric field (light grey), the magnetic field (dark grey) and the direction of propagation are all mutually perpendicular.

electromagnetic force. From the experiments of Ampère, Faraday and others it was known that the motion of an electric charge generates a magnetic force, and that the motion of a magnet induces an electric current.

Maxwell showed in his theory that this leads to the unification of the two phenomena, and predicted that therefore these two can continue to jump over each other (Fig. 1.3): a changing electric force generates a magnetic one, which in turn induces an electric force, and so on. The resulting electromagnetic waves are now familiar: for example light, radio waves and X-rays. In vacuum, all move with the same speed: the speed of light.

Once electromagnetic waves were recognized as an important constituent of our Universe, a thorough study of their behaviour was undertaken. In particular, people wondered how such waves could propagate: what was it that did the waving, so to speak? Some suggested that space was filled with an invisible kind of matter, a luminiferous aether, thereby bringing back Descartes's subtle matter in a different guise and for a different purpose. Here the problems associated with the Newtonian notion of action at a distance returned with a vengeance; the presence of an aether – even though its constituent particles, whatever they might be, could not be observed – has measurable consequences for the propagation of light. If the waves are carried by any subtle matter whatever, then their observed propagation must depend on the motion of their source (or the observer) through that aether.

Experiments to detect the effects of such an aether drift were undertaken. Soon it appeared, from the experiments of Michelson and Morley in 1887, that there is no apparent motion at all with respect to an aether; they found that the speed of light does not depend on the state of motion of emitter or observer, but that it is the same for everyone, regardless of their behaviour.

This is often expressed by saying that *the speed c of light is invariant.* The invariance of *c* is a certified observational fact, and it is surely very odd; how odd, can be glimpsed as follows. Light behaves, at least in part, like a wave. Now from daily experience we know that the speed of a wave depends on how we move. When a wave rolls towards the shore, you see the crests move at a certain speed; but a seagull can adjust its airspeed in such a way that the wave seems to be standing still below it. Now I can ask: do light waves behave like that?

Michelson and Morley took a light beam, split it in two with a half-silvered mirror, and sent the two half-beams along on different tracks, perpendicular to each other. One path was placed along the direction of Earth's orbital motion, and the other at right-angles to it. After the light rays had traversed their paths, they were brought together again, and their travel times were compared by means of an optical effect called *interference* (we will see much more of that later).

The results showed conclusively that, whatever you try, the travel time of light in vacuum along equal distances is always exactly the same. In other words, *the speed of light does not depend on the motion of the emitter or the receiver* of the light.

In daily life you won't notice this bizarre fact or its bizarre consequences. The reason that you do not observe anything particularly odd when you are cycling along is that you move so very, very slowly. No insult intended, but as far as light is concerned (which travels at 300 million metres per second) you are almost standing still. It is this, and nothing else, that makes the consequences of the invariance of the speed of light so contrary to intuition.

Notice, by the way, that the fact that *c* is invariant provides a standard of speed for the entire Universe! If the difference in speed between two objects were only a relative thing, it would make no sense to distinguish between fast and slow. But, contrary to popular opinion, it is false to say that 'everything is relative'. Because of the invariance of the speed of light, it is more nearly correct to say that everything is absolute!

This observed invariance is a most curious fact; not so much because we must abandon the idea of an aether, but rather because it leads to an amazing conclusion about motion in general. The remarkable consequences of the constancy of the speed of light can be qualitatively understood by asking: 'If c is really invariant, then how does one light ray see another one move?' Well, if there is no conceivable superseagull that can fly in such a way that a light wave appears to be standing still around it, the answer must be: 'Any light ray sees any other one move with the speed c.'

If a light ray, travelling at speed c, encounters another one head-on, also travelling at c, then the above can be written symbolically (not algebraically!) as "$c+c = c$." Similarly, if two light rays that travel in the same direction see each other move with the speed c, we must, in some sense, likewise require that $c - c = c$. If it is really true that c is invariant, then we cannot escape the conclusion that $c + c = c - c = c$! *Most* remarkable: one would have expected $2c$, or 0, or something in between, depending on circumstances. But the Michelson–Morley experiment has shown that the speed c does *not* depend on circumstances.

Einstein, who first asked the question about the relative motion of light rays, had the courage and the insight not to reject $c + c = c$ out of hand as absurd; all it means is that the invariance of the speed of light compels us to accept that speed is a more complicated beast than we suspected. However, the implications are staggering: we can, in fact, add speeds in such a seemingly contradictory way, but only at the expense of a drastic revision of the classical concepts of space and time.

Because it is required that $c + c = c - c = c$, speed cannot be a simple algebraic number for which the normal rules of addition and subtraction hold, as in the case of money or apples in the market. We must reconsider what exactly is meant by the addition of speeds. Note that this is no cause for dismay; even in everyday experience, we deal with quantities for which two plus two does not always equal four. Your position on Earth is a case in point: if you walk two kilometres, then another two, and then two more, you are not necessarily six kilometres from your point of departure. In fact, if you have walked along an equilateral triangle, you are back where you started. If distances added like money, you could never mail a letter or walk the dog: you would never get back home.

1.3 Relativity and fields

Einstein showed how we ought to define addition in such a way that the addition of any two velocities leaves the speed of light invariant. In order to be able to do this, he had to abandon the idea that time and space can be measured independently: if speeds add in a curious way, then this is due to the underlying behaviour of the ingredients of speed, namely space and time (remember, speed = distance/time!) If space and time cannot be measured independently of one another, then there is no such thing as universal time.

Subsequently, Einstein found a number of other results that are, to our intuition conditioned by always moving much slower than light about as bizarre as $c + c = c$. One of these results is that c is an absolute *maximum* speed: nothing can travel faster than light. This conclusion can be glimpsed from the above. If the addition of speeds is defined such that $c + c = c$, then we must also have $c + c + c = c$, and so forth: we can never exceed the speed of light. Notice, by the way, that I have hereby shown that c must be an upper limit because of the observed fact that the speed of light is always the same; I do *not* say that c is the maximum speed because I have tried hard to exceed it and have failed! The c limit is an inescapable consequence of the experimental fact of c invariance, and so we can prove that any attempt to exceed the speed of light must fail. Because c is finite and maximal, and because it is the same no matter what you do, it serves as a universal standard of speed. There is an *absolute* meaning to the expressions fast and slow: motion with a speed that is much smaller than the speed of light is slow, any other motion is fast.

Another result is that mass and energy are essentially the same. This, too, can be appreciated on the basis of the above. I will show this by means of a `space-time diagram`, which is a graphical summary of 'where is what when' in a particular case (Fig. 1.4).

Suppose that we have a particle that has no internal structure, and that is acted upon by one force only. We know that the more energy is transferred by the force to the particle, the faster it travels. But because c is the maximum speed, I can transfer all the energy I want, I will never exceed the speed of light. Then where did all that energy go? Recall from everyday mechanics that the transfer of a given amount of energy gives a large boost to the speed of a low-mass object, and a small speed increment to one with a high mass. Let us look at the behaviour of a football and a railway engine (Fig. 1.5) when each is given a standard kick, for example one delivered by the goalkeeper of the Dutch national team.

The ball is propelled to a high speed, which in a space-time diagram means that it lies on a line that is inclined strongly with respect to the time axis. But the same kick, delivered to a railway engine, has almost no effect; even if all eleven team members were to hurl themselves at the machine, its speed would barely differ from zero, and its space-time path would be practically parallel to the time axis. We summarize these experimental facts by saying that a football has a small mass and a railway engine has a large mass.

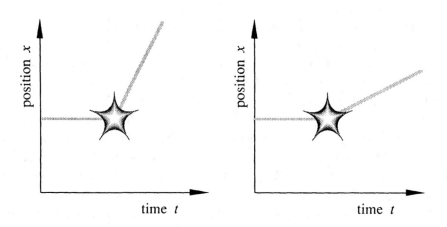

Fig. 1.4 Sequence of events of two different objects during a kick, as shown by their tracks in space-time. Such space-time diagrams occur frequently in this book, but the axes will usually be omitted. The convention I will use is that time runs to the right and spatial distance increases upwards on the page. This is different from what you usually find in the physics literature: there, time runs upwards on the page. That convention is typographically inept and I will almost never use it in this book.

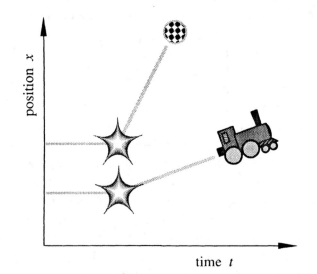

Fig. 1.5 Space-time diagrams of two different objects during a kick. The top track shows a football being kicked; its low mass shows up in a large change in velocity (i.e. a large acceleration, producing a sharp kink in the space-time track). The other track is not as sharply deflected, showing what happens when the same kick is delivered to a railway engine.

Now if c is to be a strict upper limit, the space-time paths of objects cannot be arbitrarily steep. If the football has a small speed, the kick delivered by our goalie makes a noticeable bend in its space-time path. But if the ball moves close to c, the same kick must make a much smaller bend: because c is a strict upper limit, the closer we get to the speed of light, the smaller the increase of the velocity becomes for a given addition of energy. Therefore, the more energy a football has, the more it resembles a railway engine: its energy acts as if it were mass! Because the c limit forces us to accept that energy is essentially the same as mass, we must also accept that even the mass m of an object at rest is equivalent to a certain quantity of energy E. A precise calculation of this equivalence leads to the famous $E = mc^2$.

These consequences of the experimental fact that the speed of light is the same for all observers are described precisely in the theory of relativity. In our discussion of forces, the facts of relativity have a very profound influence, mostly because of the findings that: first, the speed of light cannot be exceeded; second, mass is equivalent to energy; third, because a universal time does not exist, the order in time of events can be different for different observers; fourth, the speed of light can be kept constant only if we treat space and time on an equal footing. These points will be discussed in more detail later.

The facts and conclusions of relativity slash all support from under the Newtonian concept of instantaneous action at a distance. Because the speed of light is a maximum, there is no such thing as an instantaneous connection between spatially separated points. Accordingly, if there is to be any influence across a spatial distance, we must accept that that influence is underway for a while. Relativity prevents a force from acting instantaneously, so that there must be something that transmits it, some sort of messenger substance that carries the information about the action of the force from one point to another. This something is called a field, and because of the c limit we are compelled to accept the field as a physically real object, not merely as an aid to calculation. It is beginning to look after all as if we need some sort of direct contact to transmit a force (at this point, Descartes smiles).

Fields can have different forms, from very simple to very complex. We may imagine a field as follows: at every instant in time, each point in space is provided with a little label, on which we can read the strength of the force. When the field is simple, the label contains just one number ('scalar' field; Fig. 1.6). In a more complex case, the tag contains more numbers.

On this two-dimensional sheet of paper, I can represent a scalar field by a greyscale: the darker the picture, the stronger the field, and a single number (the percentage of paper covered by ink) describes the field at each

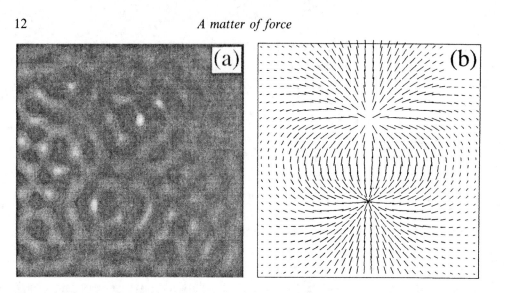

Fig. 1.6 (a) A scalar field. The field amplitude is indicated by a single quantity which, in this diagram, is represented by the intensity of the grey shading, as on a black-and-white photograph: the blacker the print, the stronger the field. (b) A vector field. Here, the field amplitude has two components: a strength and a two-dimensional direction. The strength is indicated by the length of the line segments, the direction by their orientation.

point. A somewhat more complicated field requires more numbers (called components) to describe it at each point. In two dimensions, I could represent a two-component field by a colour scale, where each component is shown as the percentage of paper covered by a different ink (say, blue for one component and yellow for the other – this would allow us to speak truly of 'green fields'). In practice, it is much clearer to represent the field at a given point by means of an arrow, which is why this type is called a *vector field*. The base of the arrow is the point where the field is measured; the directional angle is the first field component, and the arrow's length represents the second component. Thus, we obtain a field of arrows.

Complicated fields are more difficult to represent graphically, so that we usually prefer to work with an array of numbers that indicate the values of the various field components. Notice that we have a certain amount of leeway as to how we choose the components; for example, instead of taking the direction and the length of an arrow as the field components, we could have taken the lengths of the projections of the arrow in two different directions.

Fields are like weather maps. The temperature at ground level is a scalar field: single numbers all over the map suffice to specify it. The wind velocities form a vector field: notice the little arrows on weather maps. The wind shear is a still more complicated field, and cannot be represented easily on a map. I'm sure that airplane pilots wish it could!

If we require the fields to behave properly according to the laws of relativity, it turns out that we are obliged to arrange their components in powers of four: a single number (four to the power zero, `scalar field`), a row of 4 numbers (4^1, `vector field`), a square array of 4×4 numbers (4^2, `tensor field`), a cubic array of $4 \times 4 \times 4$ numbers, and so forth. The magical number 4 comes in because of the combined $3 + 1 = 4$ dimensions of space and time.

1.4 Feynman diagrams

Our everyday experiences, limited as they are to speeds that are very much smaller than *c*, have shaped our thinking in ways that make relativistic phenomena very counter-intuitive; and now we encounter yet another oddity. Not only does relativity compel us to accept force fields as physical entities, but it turns out that the action of a force at a given space-time point cannot be varied continuously. Therefore, the above description of a field as an infinite collection of tags with numbers on them is found to be incorrect (Fig. 1.7).

Instead of having, at each point in space-time, a number that indicates the *strength* of the force, we find written on each label the *probability* that the recipient of the force gets a certain standard push (say, the kick of the goalkeeper of the Quark Quantum Football Club). The field is `quantized` and the entity that delivers the push is a `quantum`.

A quantized field is very different from a continuous one. Suppose one were to interpret the force exerted by all railway trains as a continuous field. Then we could stand between the rails anywhere in the country and be at ease: the average force is so small that one's equilibrium can be maintained by just leaning forward a little. But in reality, the force is not continuous but quantized: there is either no train at all, or a very heavy one, so that – depending on the quirks of fate – you are either completely safe or completely shattered.

So it is with fields: they are quantized, that is to say, they are built up from quanta that are exchanged between the participants in the force. The particle that emits the field quanta thus resembles a mad gunner, spraying a rain of bullets in arbitrary directions. We saw that a force is symmetric between its source and its recipient: it is a mutual affair, so that, properly speaking, we

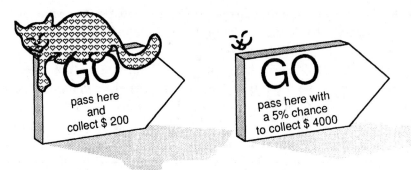

Fig. 1.7 Pictorial analogy of a classical and a quantum field. A classical field prescribes, at each point in space-time, the strength of an interaction. In a quantum field one has instead the probability that a certain interaction takes place.

should consider the emission-plus-absorption of a field quantum as the basic mechanism of a force.

Only when a quantum bullet is absorbed does anything happen. Bullets that end up nowhere† might as well not exist, and whether the target stoppeth one of three or one of 137 makes no difference. Whether you are the shooter or the one that is hit depends on the time order of events. As we saw in the preceding discussion on relativity, such time order is meaningless because it depends on your state of motion relative to the source and the recipient of the field quantum. The gunner and the target must be considered as interchangeable in a relativistic theory; accordingly, the exchange of a field quantum is truly mutual.

The fun of all this is that we are handsomely rewarded for having built two counter-intuitive concepts (relativity and quantization) into our description of a force. Because of relativity, the force between two spatially separated points must be underway for a while, and, because of quantization, the force must be transmitted by discrete quanta. As our reward for having been bold enough to include these features, we obtain a mental image that is intuitively easy: *a force arises through the exchange of field quanta.* Thus, the field quanta form a direct contact bridge between particles, so that the Newtonian concept of instantaneous action at a distance is shattered (at this point, Descartes looks as if he's about to burst out laughing).

---

† The analogy breaks down here. Unstopped bullets are forbidden by the conservation of mass and energy. We will see some curious effects of quanta being stopped by the vacuum in Chapters 12 and 13.

Fig. 1.8 Sequence of snapshots indicating the exchange of a heavy beanbag between two skaters standing on a frictionless surface.

A force appears upon the exchange of a field quantum as follows (Fig. 1.8). Let two skaters glide on a frictionless ice surface; one skater throws a beanbag at the other. This changes the velocity of the thrower; when the other skater catches the bag, the velocity of the recipient also changes.

An observer, far above the ice surface, does not see the beanbag, but does notice that the skaters change their velocities; hence the observer concludes that there is a force between them. In this analogy, we only see a repulsive force. Later we will note that the same exchange mechanism can produce attractive forces as well; in fact, repulsion is the exception and attraction is the rule.

Pictorially, the exchange event is described by a Feynman diagram. In such a diagram (Fig. 1.9) the three spatial dimensions have been collapsed into one (the vertical direction on the page). The dimension of time is represented by the horizontal direction (from left to right on the page). This layout is exactly the same as the one I used in the space-time diagrams discussing footballs and railway engines. The space-time tracks of the skaters are indicated by continuous lines, the track of the beanbag by a wavy line.

The point of the throw or the catch, where three lines come together, is called a vertex. A quantum is not a 'minimal' parcel of energy or matter or whatever. There's nothing quantized about a quantum: you can make it as big (e.g. high-energy) or as small (low-energy) as you like, and continuously to boot. It is the *interaction* which is quantized, in the sense that it is all-or-nothing. This is symbolized by the vertex in the Feynman diagrams.

You will get a clearer picture of what happens during the exchange of a field quantum if you transform the Feynman diagrams in this book into motion pictures. That can be done by making a Feynman diagram scanner

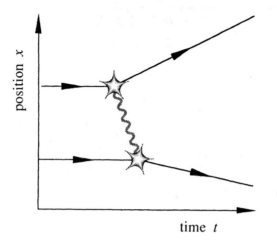

Fig. 1.9 First-order Feynman diagram. The snapshots in the preceding diagram can be derived from this space-time sequence by looking at what happens at various times. A snapshot corresponds to the situation on a vertical line in this diagram.

(Fig. 1.10). Take a piece of thin white cardboard, about 10 by 20 centimetres in size, and in the middle cut a 5 cm slot across it, with a width of a millimetre or so. Place the scanner with the slot vertically on the leftmost side of a Feynman diagram, and then slide it to the right. In the slot will appear the positions of the various quanta as time goes on. This gives a dynamic picture of what is happening. Especially with the more complex diagrams we will encounter later (involving antiparticles and all), this trick is very helpful; you are urged to scan each Feynman diagram you encounter with the FD scanner.

At this point, the analogies used for the description of a force have perhaps been stretched to the extent that you start to make objections, probably along the lines of those listed below (each of these will be discussed in detail later, but it is proper to at least mention them here).

First, why do we never see the beanbag? The reason is precisely the quantization I have invoked to describe the force. *Either you've got a quantum, or you don't; a fraction of a quantum is never observed.* If you want to see the quantum that the particles exchange, you must absorb it in its entirety; and if you do, the field quantum will not arrive at the second vertex. You may arrange to pass the beanbag along after inspection, but that would also spoil the connection between the vertices: the process you have created by intervening with your observation is not the same as the process that gave rise to the force.

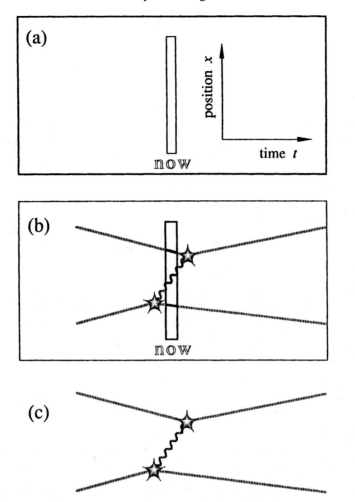

Fig. 1.10 (a) A Feynman diagram scanner, made by cutting a narrow slot in a piece of cardboard. The slot indicates the situation at the present time. By looking at the Feynman diagrams in this book through this scanner, you may produce snapshots as in the previous figures, or motion pictures by sliding the scanner in the direction of increasing time (from left to right on the page). Diagram (b) shows the scanner in action. The Feynman diagram in (c) can be used for practice.

It isn't that the intermediate quantum is unobservable; it can be observed just fine, but your observation does not give you a fair view of the quantum exchange. With quanta, it is impossible to take a careful look: either you look or you don't. Here the beanbag analogy goes grossly wrong because Nature behaves in an essentially different way on an atomic small scale than on a human large scale (we will presently see that small is in fact defined as that scale of size below which quantum behaviour becomes overt). In summary: we see the consequences of the exchange of the field quantum, but – unless we want to mess up the force – we cannot see the quantum itself. Accordingly, the latter is called a `virtual quantum`.

Second, you have probably noticed that the beanbag-throwing mechanism works for skaters, but cannot be expected to work exactly as stated for subatomic particles. An electron, say, doesn't have an internal store of energy that can provide the work done in the throw. Thus, energy cannot be conserved when a virtual quantum is emitted. That appears to be a shocking statement, and yet it does not matter: as we have seen, the force comes about through *exchange* of a virtual quantum, so we are only interested in the overall conservation of energy after the quantum has been caught by the target particle. We couldn't possibly observe whether or not energy is conserved at a single vertex, because the virtual quantum must be absorbed entirely in any process designed to measure its energy; and this, as before, would devastate the effect we were trying to observe.

Third, one might ask: what's in the bag? As it happens, this is one of the main themes in this book. The beans in the bag represent information of some sort that is exchanged between the vertices, and by putting lots of databeans in the bag we may construct forces with very complex and subtle behaviour. We can expect that the laws of relativity and quantization place restrictions on the data which the bag can carry. Moreover, we may hope to discover some general principle, over and above these laws, that prescribes what quantum beanbags may or may not contain. We will find that such a principle exists: `symmetry`. Current speculations in theoretical physics suggest that the contents of the beanbags are in fact stored in higher dimensions outside the common four of space-time (Chapter 14).

1.5 Matter and force

In the above, we saw that the laws of relativity and quantization lead us to consider exchanged field quanta as the carriers of a force. In the analogy given, the beanbag represents the force and the skaters represent the matter on which the force acts. With the powerful bias that gross everyday physics produces in our minds, it seems natural to think that matter and force are totally different things. And yet, having carefully considered how relativity and quantization led to the concept of the exchange of quanta, you may wonder why we shouldn't occasionally expect to see two beanbags throwing a skater at each other!

What is it, then, that produces the radically different behaviour of matter and force in our large-scale world? This is the thrust of the next seven chapters, but I think that it is important to discuss the distinction between matter and force briefly here. As we will see, the laws of relativity and quantization, together with certain properties of space-time, imply that all quanta can be divided in two classes: the Fermi–Dirac particles or fermions, and the Bose–Einstein particles or bosons. To which class a particle belongs depends on the amount of rotation it carries. This amount is indicated by a quantity called spin angular momentum, or spin for short. Spin is quantized. In suitably chosen units, the only values that the spin s of a quantum can assume can be written as $s = n/2$, where n is a whole number: $0, 1, 2, 3, \cdots$. If n is an *odd* number, the particle is a *fermion*; if n is *zero or even*, it is a *boson*. Thus, we have $s(\text{fermion}) = \frac{1}{2}, \frac{3}{2}, \frac{5}{2}, \cdots$, and $s(\text{boson}) = 0, 1, 2, 3, \cdots$. A list of the more common bosons and fermions is given in the table below.

name	symbol	spin	el.charge
photon	γ	1	0
weak photon	W^+, Z, W^-	1	1, 0, -1
gluon	g	1	0
electron	e	$\frac{1}{2}$	-1
neutrino	ν	$\frac{1}{2}$	0
proton	p	$\frac{1}{2}$	1
neutron	n	$\frac{1}{2}$	0

In most Feynman diagrams (Fig. 1.11) we see fermions exchange bosons, such as in the scattering of one electron off another. However, it is perfectly possible to have bosons exchange fermions, for example in photon-photon scattering.

The existence of spin is intimately associated with the fact that, in three-dimensional space, objects can be rotated about an axis that lies inside that space. For example, the rotational axis of Earth points to the star Polaris, and not towards some point outside our Universe. If Earth were a flat circle, it could still rotate, but the rotation axis would *not* lie in the two-dimensional universe of such a flat Earth: it would be perpendicular to it, and be 'out of this world'.

Because spin is the amount of rotation that a quantum carries, it is plausible to expect that the way in which a particle behaves when it is

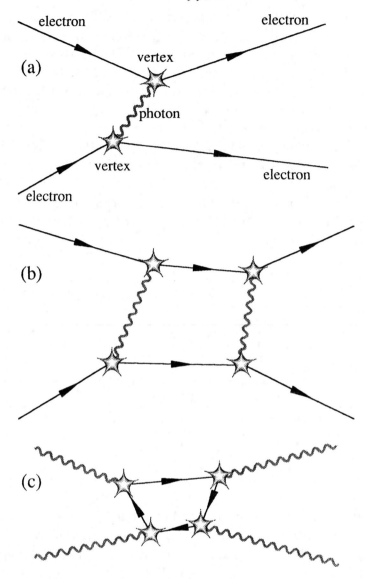

Fig. 1.11 (a) Scattering of one electron off another, shown in a Feynman diagram. Here, a single photon is exchanged, so that two vertices occur. This is a first-order diagram. (b) The same electron–electron scattering, but now two photons are exchanged (second-order diagram). (c) Scattering of one photon off another. Because two electron lines and a single photon line come together at a vertex, photon–photon scattering can never be a first-order process. An actual interaction is the sum of all its possible Feynman diagrams, so do not think that diagrams like these tell the whole story!

rotated in space is related to its spin. And indeed, it can be shown that a boson remains unchanged when it is given a full 360° rotation, but that a fermion requires *two* such twists, or 720°, to remain unchanged! If a fermion is rotated over 360°, it reverses its sign in a way that will be specified more precisely in Chapter 8.†

This sign-reversal under rotations implies that fermions and bosons behave in radically different ways when bunched together. It is found that fermions are totally intolerant particles: not more than one fermion can exist in any given quantum state. In other words, if we were to prepare a fermion in a particular state (e.g. at the North Pole, with its spin in the direction of Polaris), then it would be impossible to have another fermion in that state, or even in a state that is almost the same. Contrariwise, bosons are gregarious particles: it is more common to find two bosons in the same quantum state than it is to find them in different states.

The boson-fermion behaviour of quanta goes by various names. It is properly called 'the law of spin and statistics'. Sometimes it is called the Pauli exclusion principle when applied to fermions only (in view of this gentleman's legendary crabbiness, 'Pauli intolerance' would be a more appropriate term).

The difference between intolerant fermions and gregarious bosons produces the apparent difference between matter and force. First, consider an individual fermion gunner who sends off a spray of bosons. Because of the gregarious behaviour of bosons, they can coexist when shot off in closely similar directions. Indeed, in any given process bosons are more likely to be in identical states. Thus, bosons exchanged between fermions can build up a coherent quantum field that can, in principle, extend over long distances. Contrariwise, an individual boson that sends off a spray of fermions does *not* build up such a field, because the intolerance of the fermions implies that they can *not* coexist when shot off in closely similar directions.

Second, consider a collection of fermions. Because of their intolerance, they must all be in different quantum states. Thus, it is impossible for identical fermions to all assume the state of lowest energy, and collapse: a collection of fermions will always form a lump of finite size. But a collection of bosons will assemble in the lowest energy state they can reach. In principle, you can put an arbitrarily large amount of light (photons are bosons!) into a box, but only a finite quantity of fermions (e.g. electrons) will allow themselves to be squeezed in. Matter and force are not fundamentally

---

† See R.P. Feynman's contribution in *Elementary Particles and the Laws of Physics*. The reversal of the sign applies to the amplitude of the process in which the fermion takes part; it does not mean that it turns into an antiparticle, or anything like that.

different, but are the names that we have given (guided by mere large-scale behaviour) to collections of fermions and bosons.

In summary, the two classes of particles show the following behaviour: fermions act like infants. You always have to keep them out of each other's range. The one in the sandbox won't tolerate number two; you must put it in a stroller. That won't hold more, so the third infant must ride a bicycle. Number four takes a motorbike, number five a car, and so forth, all the way up to the speed of light. Thus fermions, like infants, take up an amount of space that is gigantic compared with their size. Bosons are quite the opposite. They behave rather like rugby players. As soon as one hits the ground, both fifteens pile themselves right on top, forming a wriggling heap. These are chummy particles which squeeze themselves in the smallest possible volume, never taking up much space but capable of acting as a coherent team.

Fermions collect in definite lumps, but cannot produce a coherent field; a batch of bosons collapses without much resistance, but can act coherently when exchanged. Therefore, fermions appear to us as matter, whereas bosons provide what we call force. Thus, I was *very* amused to see the clash of light sabres in the *Star Wars* film trilogy: since a light sabre is presumably made of particles of light (which are bosons) it will hardly stand up to impact! The law of spin and statistics gives an excellent reason to make swords out of fermions (such as atoms of iron and carbon). Our understanding of Nature involves two closely related quests: the search for the free fermions that occur (often loosely called fundamental particles), and the search for the bosons that they can exchange (occasionally called field particles).

2

Stalking the wild rainbow

✪

2.1 Colours and spectra

There is a striking similarity between the struggle of today's physicists with particles and the pursuit by their predecessors of a hot topic: spectroscopy. A lot of the jargon of the present relativistic quantum field theories comes directly from spectroscopic descriptions. Our search for order in the bewildering array of particle masses is like the efforts of scientists who, towards the end of the preceding century and in the first decades of the the twentieth, tried to find some order in the arrays of light waves that can be emitted by atoms.

It has been known since time immemorial that there are colours, but it wasn't until the seventeenth century that it became clear that all colours are different manifestations of the same phenomenon: light. The behaviour of light began to yield to quantitative descriptions through the brilliant work of Snell and Huygens. The colours and behaviour of the rainbow, first correctly explained by Descartes, were then no longer a religious mystery. Newton worked systematically on the splitting of sunlight by glass prisms (Fig. 2.1) into its coloured components, and he showed that the colours can be recombined to yield white light.

Huygens's wave theory of light reached its apotheosis in the work of Fresnel, who proved that the colour of a light wave is determined by but one quantity, the wavelength. Fresnel's work also paved the way for a new gadget with which white light can be split into its constituent colours: the `diffraction grating`. Because few things are so convincing as experiments done by oneself, we will take some time out to construct a device that splits light into colours: a `spectroscope`. Seriously: with some cardboard and a small diffraction grating (Fig. 2.2), you will be doing

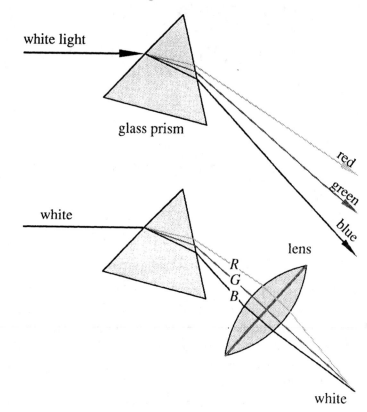

Fig. 2.1 Two basic experiments with a prism. When white light passes through a glass prism, the light is split into a band of colours. Each colour corresponds to a certain wavelength. When the colours are combined by means of a lens, the eye sees the result as white light again.

with light waves the exact same thing that giant mass analysers do in the world's biggest particle accelerators.

A diffraction grating is a transparent or reflecting piece of material, on the surface of which are cut a large number of thin, parallel grooves. Such gratings can be bought at many science museums and suppliers of scientific equipment. A grating is a multimillion-aperture version of the double-aperture diffraction experiment which we will discuss in detail in Chapter 5. Light passing through the grating is deflected by a process called diffraction, and the amount of the deflection is proportional to the wavelength of the radiation. Thus, a diffraction grating sorts light waves by wavelength.

Now take a piece of grating, hold it close to your eye, and look through it at a small incandescent light bulb in the distance. You will see the bulb, with

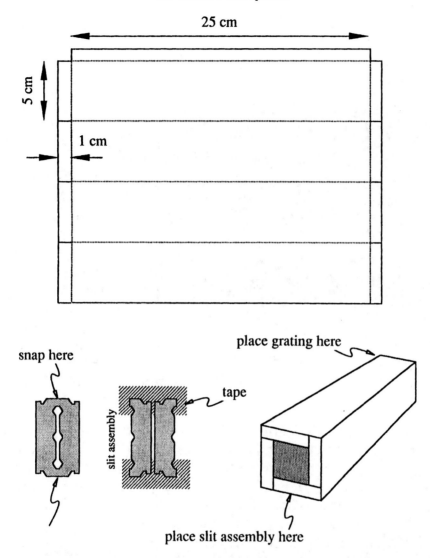

Fig. 2.2 Plan of a spectroscope box. Cut the pattern shown here out of cardboard (preferably black). Assemble it to make a long box. Make a narrow slit from the two halves of a razor blade (*be careful!*) or from two straight strips of folded aluminium foil. Place the slit over one end. On the other end, place a plastic diffraction grating, as sold by various science supply shops, or in some science museums. Make sure that the grating is aligned properly with the slit.

on each side an image of it, smeared out into the colours of the rainbow. This band is called the spectrum of the light. We can improve on this experiment by constructing a box for the grating which blocks stray light red   orange   yellow   green   blue   violet

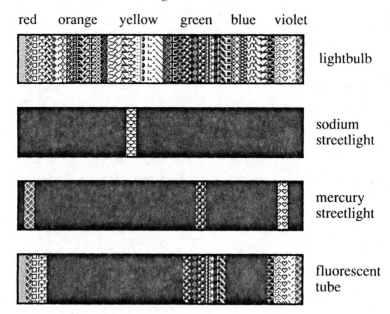

lightbulb

sodium streetlight

mercury streetlight

fluorescent tube

Fig. 2.3 Sketches of continuous and line spectra. Red is on the left, blue and violet on the right. The funny textures represent the various colours, which are irreproducible in a black and white book. All the more reason to do the experiment!

from the surroundings. Moreover, the box can be fitted with an entrance slit to sharpen the image of the spectrum. The slit and the box effectively act like a pinhole camera, but here the pinhole is small only in one direction and elongated in the other. Fix the slit on to the box, and rotate the grating until the spectrum is broadest. The grooves of the grating are then parallel to the slit. Keep the grating close to your eye, and look at a light source through the slit at the other end of the box.

2.2 Spectral lines

Now look at as many different light sources as you can find. In particular, look at the neon signs in shop windows and at sodium or mercury city lights (see Fig. 2.3 for a rough sketch of what to expect). You will notice that the light from many of these sources is not a continuous band but contains a number of individual streaks of pure colour. These are called `spectral lines`. Investigate the lines from different sources. Are there regularities, similarities, differences?

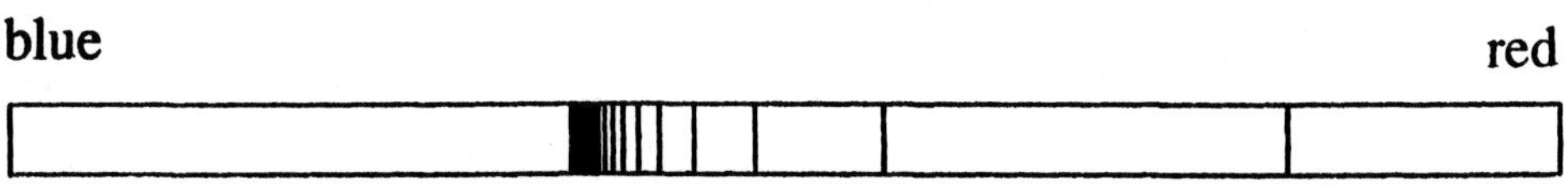

blue   red

Fig. 2.4 The Balmer series of spectral lines due to the hydrogen atom. Blue is on the left, red is on the right of the diagram.

When physicists started to split white light into colours, it was soon noticed that there are occasions when the colour band is not broad and diffuse, ranging from red to violet (as in the Sun or in a candle flame), but, instead, shows light at only a few sharply defined colours. For example, if you look at a candle flame with your spectroscope, you will see the familiar rainbow band, but if you sprinkle some salt in the flame you will temporarily see a brighter light in the yellow part of the spectrum. Because practically all spectrographs are fitted with an entrance slit, and because the light we observe in the spectrum is an image of that slit, such an isolated colour appears as a transverse line in the spectrum, a spectral line.

Soon it was discovered that every chemical element has a unique set of spectral lines associated with it (Fig. 2.4). Also, it was seen that the lines are not randomly distributed over the spectrum. On the contrary, a typical line spectrum consists of many superposed families, called `series`, of lines which increase in wavelength in an orderly progression. Often, progressions were seen of small bunches of lines which somehow belong together; these were called `multiplets`. For lack of anything better to do, spectroscopists gave fancy names to spectral series; some were called 'sharp' (abbreviated S), some 'principal' (P), 'diffuse' (D), 'fundamental' (F), and so forth. Clearly, these were heavily prejudiced indications, based on nothing but phenomenology; we now know that there is absolutely nothing principal or fundamental about the P and the F series. It will be seen later that S, P, D, F, and so forth, refer to the spectral series that arise from different quantum states† of an atom. But the SPDF designation is still being used, even in the quantum theory of the scattering of particles. Who knows what future generations will think of the fact that we have invented the names 'up', 'down', 'charm', 'strange', and so forth, for certain abstract properties of the particles we call quarks! Our only defence is that we do realize that these are merely arbitrary and superficial labels.

The historical effort to assign spectral lines to orderly arrangements (Fig. 2.5), or at least to groups that closely resemble each other, is analo-

† Specifically, states with different *angular momentum.*

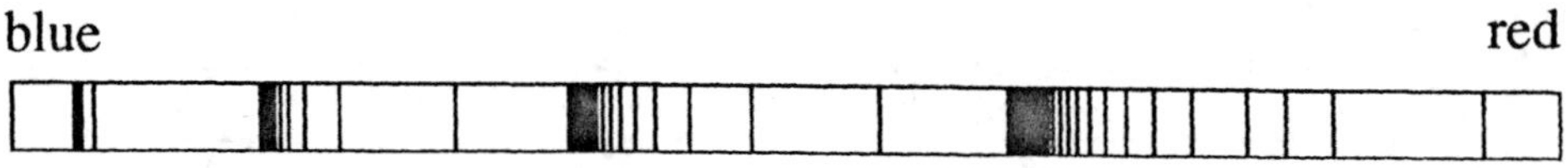

blue   red

Fig. 2.5 Hydrogen spectrum as built up from the Lyman, Balmer, Paschen and Brackett series. Blue is on the left, red is on the right of the diagram. An infinity of such series exist, but these are the strongest and most easily observed.

gous to the contemporary search for 'fundamental multiplets' of fermions. For example, the spectrum of hydrogen begins to look a lot more comprehensible if it is seen as overlapping families of spectral series. Likewise, the three most familiar particles show a pattern that suggests classification: the lightweight electron on the one hand, and the much more massive proton and neutron on the other. In this view, the electron is a representative of a family of light particles (leptons), whereas the other two belong to the heavy-particle family of the baryons. Closer inspection shows that the proton mass is only 0.14% less than the mass of the neutron, suggesting a very close relationship between these two, to which we will return in Chapter 10.

2.3 Classical physics stumped

It proved to be utterly impossible to explain the spectral lines by means of the Newtonian physics of the nineteenth century. One of the worst problems can be appreciated by analogy with a classical mechanical oscillator such as a guitar. If you pluck a string very gently in the middle, you will hear a fairly pure tone. But if you pluck it strongly, or near the bridge of the instrument, the tone is distorted. This phenomenon occurs in every classical oscillator: besides its pure frequency, called fundamental, it also produces harmonics (Fig. 2.6), such as those which musicians call the third, fourth, fifth, octave, and so on.

Each line in an atomic spectrum occurs at a fixed frequency, which our eyes see as a colour. However, the expected harmonics are *never* observed. Instead, the frequencies were found to be related by a rule called Ritz's combination law. According to this rule, if f_1 and f_2 are frequencies of spectral lines, then there are also lines with frequencies $f_1 + f_2$ and $f_1 - f_2$. In this way, spectra could be built up from a fairly small number of basic frequencies, called terms. The classical theory of oscillators cannot explain this peculiar behaviour.

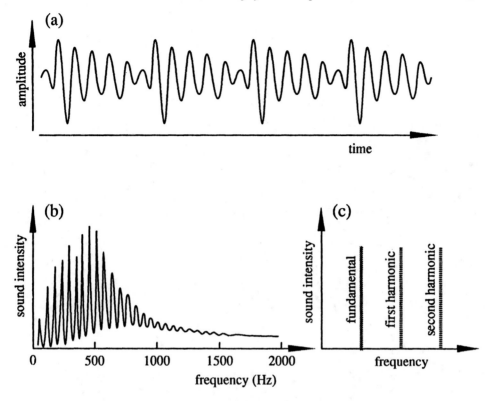

Fig. 2.6 (a) The wave form of a French horn is not a pure oscillation but contains many higher harmonics, also called overtones. These are shown in the spectrum of the sound (b); the curve gives the strength of the sound at any particular frequency. The unit of frequency is the hertz (Hz), the number of oscillations per second. (c) An indication of the place of a series of overtones in a schematic sound spectrum.

It was discovered that the differences in frequency correspond to differences in energy of the atom. This energy (Fig. 2.7) is proportional to the frequency of the emitted light. Thus, the various spectral lines could be attributed to transitions inside atoms, due to some (then mysterious) internal degrees of freedom of the atom which somehow fail to obey the rules of classical oscillators such as guitar strings.

One could try to resort to extremes in a desperate effort to rescue classical mechanics, as follows. Suppose that the atomic laws of motion are such that only exactly pure oscillations occur, without any harmonics whatever. In this view, each line corresponds to a 'fundamental' or 'elementary' light wave, unrelated to the others. This is bizarre, but not strictly forbidden, and even though it does not explain Ritz's law we may hope to get away with postulating one perfect oscillator for each spectral line. But there is a problem with this. If we shake an oscillator around, it begins to vibrate;

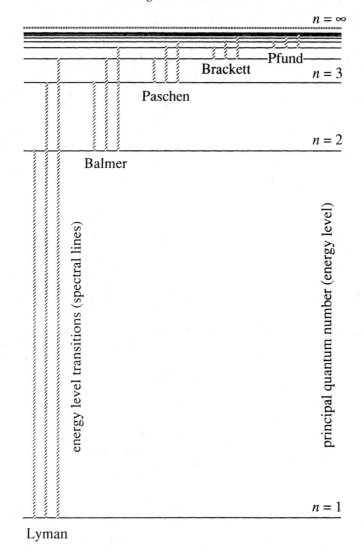

Fig. 2.7 The energy levels of the hydrogen atom. The higher the principal quantum number n, the higher the energy. Some transitions between energy levels are indicated. These are responsible for the spectral series shown in Fig. 2.5.

some of the energy used in the shaking is absorbed by the vibrational motion. Thus, if each spectral line is due to an independent oscillator, each one soaks up a little energy. But the number of lines observed in a given species of atom is enormous (in fact, it turns out to be infinite). If each line oscillator is an energy sink, we would expect to be able to add very large amounts of energy to matter without it getting appreciably hotter. This is not observed. Somehow, the oscillators corresponding to the spectral lines do not soak up energy. This is in blatant contradiction with classical mechanics.

3

Light

3.1 Waves of light

Light behaves like a wave. In the laboratory, we can measure the vibrations that are set up in matter when a light wave comes by. We can also observe the peculiar light-and-dark patterns that occur when two light rays are made to act simultaneously, an unmistakable sign of wave behaviour.

Let us make a small excursion into wave motion. Throughout this exposition, you are encouraged to experiment as much as possible with waves (preferably real ones, or at least those in the diagrams), to become familiar with their fascinating behaviour. Water waves in the bathtub, or those seen on open water from a high tower or an airplane, are especially instructive.

Waves are periodic; if you pick a point in space (e.g. the surface of a pond at the point where a reed pokes through) and you watch the motion of the wave carrier (the water that bobs up and down), then you will see that the state of motion of the carrier repeats itself at regular intervals of time. This interval is the period of the wave. A free wave has a velocity, i.e. a direction and a speed. At a given point, a wave alternates periodically between a certain maximum and a minimum. Half the height between wave crest and wave trough is called the amplitude of a wave (Fig. 3.1). The distance from crest to crest is the wavelength, and the number of wave crests going by a fixed point in one second is the frequency of the wave.

The extent to which a wave has completed any oscillation cycle (Fig. 3.2) is called the phase. For example, we may (arbitrarily) start counting at a wave crest, and call that phase zero; when the phase is half a cycle, there will be a wave trough. The phase is often expressed in degrees, so that 360° corresponds to one full oscillation cycle. The reason is that uniform circular motion can be used to generate a wave (Fig. 3.3). Imagine a series of circular wheels, each with a dot painted on its circumference. If these

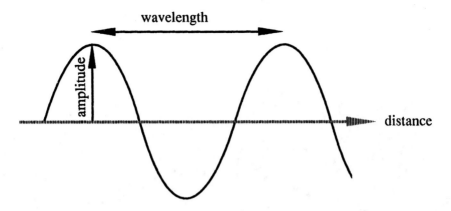

Fig. 3.1 Two of the main properties of a wave: the wavelength and the amplitude.

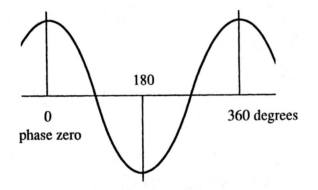

Fig. 3.2 The phase of a wave. Starting at a fixed but otherwise arbitrary point (here the point of maximum height), the phase measures the extent to which a wave has completed an oscillation cycle.

rotate in unison, we perceive the motion of the dots as a progressing wave, provided that the dot on each wheel has a constant offset (called phase shift) with respect to those on adjacent wheels. The undulating curves generated by dots on revolving circles are called *sine waves*. A *cosine wave* is the same as a sine wave, but with a phase shift of −90°. In a later chapter, we will consider the possible paths that particles can take, and we will see how the wave properties produced by a rolling wheel allow us to describe the world on an atomic scale.

A wave is not a material thing. We clearly see ripples propagate along the water surface, but closer inspection shows that a cork is not moving

Light

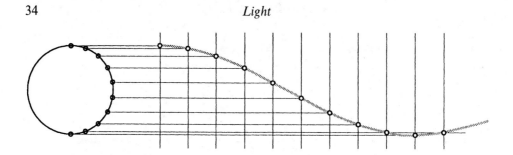

Fig. 3.3 Wave generated by a revolving wheel. Given a wheel which rotates with constant speed, one can construct the wave by taking a fixed point on the wheel and plotting its height against time. The resulting mathematical curve is called a *sine*. A *cosine* wave is exactly the same, but shifted to the left by a phase difference of 90 degrees, such that $\sin(\theta) = \cos(\theta - 90°)$.

with the wave: it merely gyrates about a fixed spot. Thus, this type of wave is a *collective state of motion* of an underlying carrier, in this case the water. Beware: water waves, sound and other classical wave phenomena are used here strictly as a metaphor for the wave behaviour of quanta. When describing quanta and their interaction, we can use the concepts and mathematics that were developed for waves in air and on water. But that does not mean that quanta are ripples on an invisible cosmic pond! Thus, we ought strictly to say 'quanta show wave behaviour', rather than 'quanta are waves'.

3.2 Huygens's principle

The way in which a wave propagates can be found by using Huygens's principle: *each point of the leading wave crest is a source of a spherical wavelet with the same frequency and phase*. Although we will come back to this subject in detail, it is worth pointing out here that the most characteristic property of a wave is that it has a *phase*. It is precisely this property that causes the rich and surprising patterns that waves can produce. To see how phasing is connected with Huygens's principle, consider a strip of material (Fig. 3.4) with regularly alternating white and black bars.

This 'phase tape' represents the regularly cycling phase along the path of a wave. If the wave passes through an aperture that is very small compared with the wavelength, the wave shape after the passage is strictly determined. No matter in which direction we lay out the phase tape, it always pivots at the aperture. This produces a spherical outgoing wave, which is the basis of the Huygens construction (Fig. 3.5).

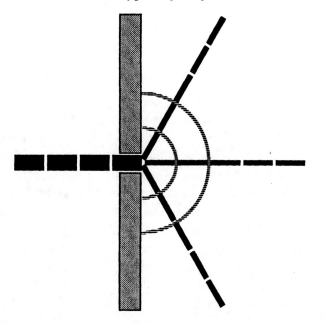

Fig. 3.4 Phase tape construction of a wave passing through a narrow opening from left to right. When the opening is small compared with the wavelength, the phase of the wave on the far side of the wall is strictly determined everywhere. This produces a Huygens spherical wavelet beyond the opening.

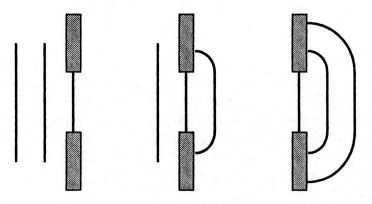

Fig. 3.5 Huygens's principle at a finite aperture. When the opening is not small compared with the wavelength, the Huygens wavelets add up along the aperture, producing a 'flat piece' in the outgoing wave.

We will assume that the speed of the wave is the same everywhere; the consequences of differences in speed can also be studied by means of Huygens's principle,† and the effects of refraction, for example – as expressed in Snell's Law – can thus be calculated; but we don't need any of that here. Suppose that the wave has reached a certain point in space. In the next wave period, each point on the wave crest emits a ripple with a radius equal to the wavelength. All those ripples together generate the next leading crest, and in this way the wave front advances. You can easily see from this construction that the direction of the velocity of a wave is always perpendicular to the wave front, so that a straight wave front remains straight. This type is called a `plane wave`. Also, a spherical wave remains spherical (Fig. 3.6). An irregular wave smooths itself out.

If a wave – say it is a plane wave – encounters an obstacle, Huygens's principle shows that the wave front curls around it. This is called `diffraction`; it explains why we can hear sounds from the other side of a building, for the diffraction allows the waves to sneak around a corner, even in the absence of reflection. You are urged to try out Huygens's principle in a variety of situations; suggestions are given in Fig. 3.7 and Fig. 3.8. An especially important case is a wave impinging on a wall with a small hole in it (Fig. 3.5). Please convince yourself by applying the Huygens construction that the wave beyond the wall is spherical if the hole is small compared with the wavelength, and that it becomes less and less spherical if the hole is made bigger. This is *very* important, as will be seen later.

3.3 Interference

In what follows, we will mostly be interested in the ways in which waves can be added together, because the resulting effects are the hallmark of wave motion. Moreover, we will see that the way in which Nature adds quantum processes corresponds exactly with the addition of waves. Consider a long train of wave crests and troughs, going from left to right. Now take a wave that is precisely the same, except that it travels at a slight upward slant. What do these waves look like if we combine them? That depends on what is meant by 'combine'.

---

† Huygens's principle is not a fundamental 'law of nature', or anything of that sort. Rather, it is a geometrically and intuitively easy rule (which can be rigorously derived from the equation of wave propagation) that summarizes the most important aspects of the motion of waves. Use of this principle allows us to describe many of the concepts of wave mechanics without having to go through heavy mathematics. It is a remarkable reflection on the genius of Huygens that he found this powerful principle long before the wave equation (a partial differential equation that strictly governs wave motion) was discovered!

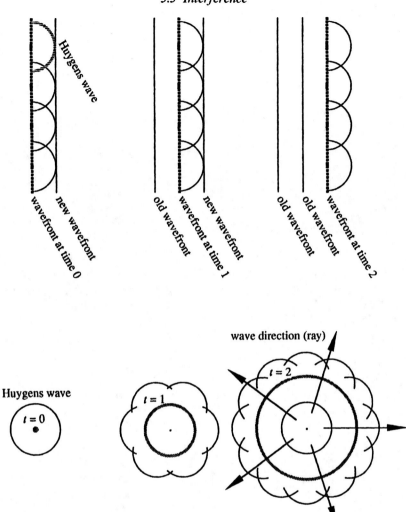

Fig. 3.6 Propagation of a plane and a spherical wave according to the Huygens construction. Each new wave front is found from the previous one by letting small wavelets propagate from each point on the old front. The direction of the wave is everywhere perpendicular to the wave front. Lines which trace these directions are called rays.

As it happens, classical waves of which the amplitude is very small compared with the wavelength (so-called `linear waves`) can be combined by simply adding the wave heights together (Fig. 3.9). This is called linear superposition, or `superposition` for short. Immediately we see that the superposition of waves generates peculiar patterns, because in some places the wave crests coincide with other crests (leading to a doubling

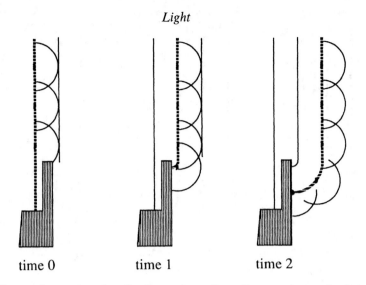

time 0 time 1 time 2

Fig. 3.7 Huygens's construction for the propagation of a wave around a flat obstacle. Notice how the construction predicts the bending of the wave around the corner; this is called diffraction. Curiously, Huygens never appears to have used this remarkable insight explicitly.

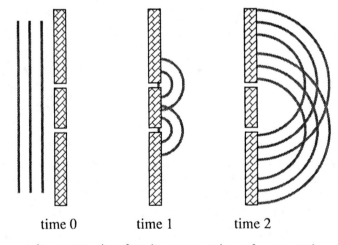

time 0 time 1 time 2

Fig. 3.8 Huygens's construction for the propagation of a wave through a screen with two holes. Notice the appearance of a zone where the two waves on the far side of the screen intersect; this is called interference.

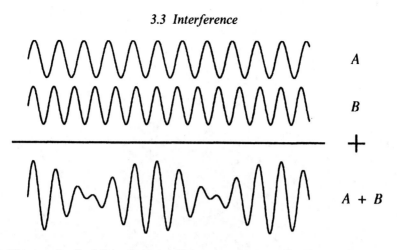

Fig. 3.9 The result of adding two one-dimensional waves. In places where two wave crests coincide, the resulting amplitude is large (constructive interference). Where a crest of one wave lies on top of a trough in the other, the result is zero (destructive interference). The total wave pattern shows a regular alternation of high and low points of the wave envelope, called beats.

of the amplitude), whereas in some other places the crests coincide with the troughs (leading to a cancellation of the motion and hence to a zero amplitude). This pattern generation is called `interference`. When the amplitudes add, we have constructive interference; when they cancel it is destructive.

An important example of interference, which we will use a great deal below, is the one in which a wave falls on to a barrier with two holes in it (Fig. 3.10). We will assume that the holes are small compared with the wavelength, and Huygens's principle shows that the wave beyond the barrier consists of two spherical waves. By superposition of these waves, we see clear interference bands, called `fringes`. If we were to place a row of observers beyond the barrier, they would report regions of double amplitude alternating with zero amplitude.

You should make a transparent photocopy of the spherical wave in Fig. 3.11, and experiment by superposing it on to its original.† You will notice immediately that the spacing of the interference fringes in the two-hole diffraction changes dramatically as the distance between the holes changes: the smaller the distance between the holes, the larger the spacing between

† Technically, this superposition is not quite the same as the one discussed above, because in the troughs the wave does not have a negative value. It is impossible to print with negative ink, so the superposition occurs by adding according to what computer buffs call *logical* AND *mode*: if 1 represents black and 0 white, then $0 + 1 = 1 + 0 = 1 + 1 = 1$ and $0 + 0 = 0$. Even so, the basic interference phenomenon persists, in the form of *Moiré fringes*.

Fig. 3.10 Interference between two spherical waves that emanate from small holes; this is a more detailed version of 3.8.

the fringes, and vice versa (Fig. 3.12). Hence we conclude that interference between two waves depends on the relative position in space of the wave sources.

Further experimentation shows that interference also depends on the relative position in time of the waves. If we arrange things in such a way that the holes in the screen show exactly the same motion (when a wave crest passes one hole, the other one lets through a wave crest also, at exactly the same time), then we say that the waves are *in phase*; the phase indicates the relative position in time of the waves, because it shows which fraction of a cycle the wave has completed. It is often useful to express the difference in phase between two waves in degrees, from 0° to 360°, after which we start again at zero. If the phase difference is 0°, the waves are in phase and they oscillate exactly in step. Thus, crests coincide with crests and troughs with troughs: when the waves are in phase, the interference is fully constructive. If the phase difference is 180°, a crest coincides with a trough and the interference is maximally destructive.

Interference can produce zero amplitude, for example when two waves with the same amplitude arrive 180° out of phase with each other. Furthermore, it is very important to realize here that only the phase *difference* matters in

Fig. 3.11 Spherical wave suitable for copying on to a transparency. With this copy and its original you can perform all the interference experiments in the text in more detail.

superposition; the value of an individual phase is unimportant. By using the phase, together with the principle of superposition, we can give a prescription for the net result of interference. At a given point in space, the amplitude A of the resulting wave depends on the amplitudes a and b of the superposed waves and on the phase difference ϕ between them, in such a way that $A^2 = a^2 + 2ab \cos \phi + b^2$. We can see directly (Fig. 3.13) that if the waves are in phase ($\phi = 0°$), we have $\cos \phi = 1$, so that $A = a + b$; if the waves are in antiphase ($\phi = 180°$), one finds $\cos \phi = -1$ and $A = a - b$.

Please verify this prescription by looking at the preceding diagrams again. By making a transparent copy of the plane and spherical wave patterns, and sliding each copy back and forth in various orientations with respect to its original, it is easy to see that the interference pattern changes when the phase changes. This happens in all cases of interference.

3.4 Standing waves

A wave can interfere with itself in such a way that the resulting pattern remains stationary (Fig. 3.14). This type of superposition is called a standing wave. Imagine, for example, a string between two suspension points, such as a guitar or piano string. A wave travelling along the string will hit an end

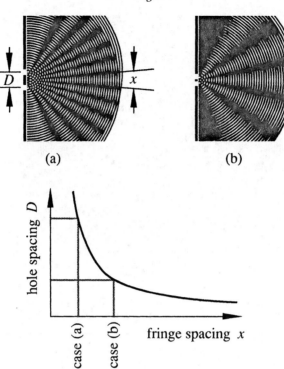

(a)   (b)

hole spacing D

case (a)   case (b)   fringe spacing x

Fig. 3.12 The effect of changing the distance D between the holes in a two-hole interference setup. In (a), the distance is larger than in (b). Consequently, the fringe spacing x is smaller in (a) than in (b). This reciprocal relation between the distance D between the holes and the fringe spacing x, shown in the graph, is responsible for the uncertainty relation of waves.

point and start moving backwards, thereby setting up an interference with itself.

If the result is stationary, we observe the sinusoidal pattern of a wave fixed in place rather than travelling along. The displacement of the string oscillates everywhere with the same period, but the amplitude changes in a regular way along the string. A position where the amplitude is zero is called a node; between two nodes the phase is constant, and adjacent regions between nodes are 180° out of phase. The suspension points, being rigidly fixed, are always nodes, and there is a fixed number of nodes for each oscillation pattern. The length L of the string is related to the wavelength λ of the standing wave by the relationship $L = n\lambda/2$. The oscillation pattern (Fig. 3.15) associated with a given value of n is called a wave mode. The quantity n is a whole number: $1, 2, 3, \cdots$. Notice how *the whole numbers arise*

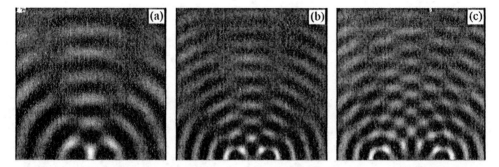

Fig. 3.13 Examples of interference for the double-hole experiment. These greyscale maps show the actual result of letting two spherical waves interfere. Notice the difference between (c), where the holes are far apart and the interference bands are close together, and (a), where the situation is the reverse. Case (b) is in between these extremes.

naturally in standing wave patterns. We already saw something very similar in the discussion of the spin s of particles, which was given by $s = n/2$. We will encounter similar cases throughout this book (e.g. in Chapter 9) as `quantum numbers`.

When a light beam from a point source is thrown on to a screen with two very small holes, the intensity of the light received on a film beyond the screen follows exactly the sequence of interference maxima and minima which we expect on the basis of the above diagrams. It is evident from this and other such experiments that we can describe light as a wave with a definite velocity, wavelength and frequency. Moreover, light waves can be combined by means of linear superposition. Radio waves have low frequency, light is intermediate, X-rays and gamma rays have high frequencies. Our eyes see the frequency of light as *colour*: red at the low frequency end, via orange, yellow, green and blue, to violet at the high frequency end of the spectrum.

3.5 Particles of light

Light behaves like a particle. In the laboratory, we can measure the photoelectric effect that occurs when light particles – called `photons` – give up their energy to catapult electrons to freedom out of the metal in which they were held captive. We can also observe the scattering that occurs when photons collide with free electrons and recoil off them like billiard balls, an unmistakable sign of particle behaviour.

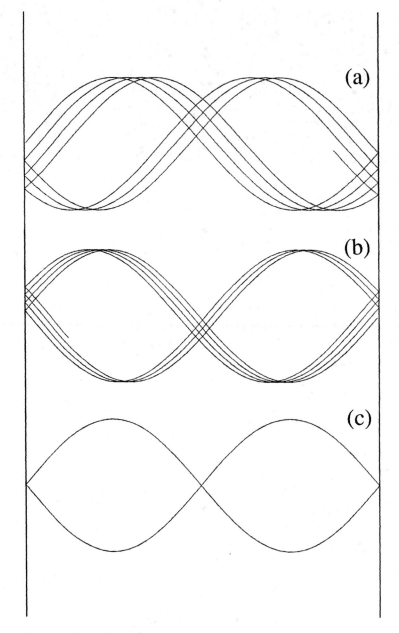

Fig. 3.14 Construction of a standing wave between two walls. In (a), the wavelength is a little too short, and the reflected wave does not return exactly to its point of departure. In (b), the wavelength is a little longer but still not long enough. Only in (c) does the wavelength match the distance between the walls. In that case, constructive interference occurs, resulting in standing wave.

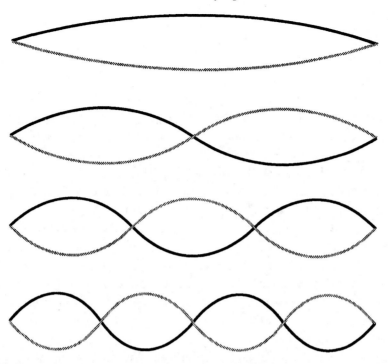

Fig. 3.15 Examples of modes on a string. The fundamental is shown at the top; below it are the successive overtones. These correspond to the harmonics shown in the sound spectrum of Fig. 2.6.

We need not make a deep excursion into particle motion. Most people are, from everyday experience, familiar with particle behaviour. But a brief reminder of some of the concepts of particle mechanics (often called 'classical mechanics') will assist understanding of many of the concepts in the book.

The basic question of mechanics is: *where is what when?* To find answers to this question, we must first find precise quantities to describe the 'where', 'what' and 'when'. In the following we will not yet talk about the effects of relativity, so we must make a mental note that changes will have to be made later.

The 'what' is a tough one to formulate; we will use the word `particle`, roughly meaning something that is an idealization of everyday objects. A particle in mechanics, sometimes called a point mass, is everything that an ordinary object like a tennis ball is, except that it does not have spatial extent. Hence, properties that require size in order to have physical consequences – shape, rotation and so forth – are suppressed.

With 'where' things are a bit easier. The quantity describing it is the `position`: an arrow pointing to the particle, starting from a suitable reference point. If we try to find a mathematical object corresponding to this distance-with-direction, it turns out that an ordinary number is not good enough. Position is a more complex beast, which doesn't obey the arithmetic laws of single numbers. As it happens, a good mathematical object with which to describe position is a row of three numbers; accordingly, we say that space has three `dimensions`.

There is a lot of leeway in the manner used to express these three numbers, called `coordinates`. For example, they might be 'forward, sideways, up', or 'latitude, longitude, height', or 'azimuth, altitude, range'. Various forms of coordinate systems have been invented to make calculations easier. Thus, we have rectangular coordinates (for Bauhaus architects), spherical coordinates (for sailors), elliptical coordinates (for navigators using Loran beacons) and so forth. But we always need three numbers to specify a position in space.

The row of numbers is called a `vector` (a vector of dimension 1, which is a single number, is called a `scalar`). We can visualize the position vector as an arrow pointing in the direction of the particle; the length of the vector corresponds to the `distance` between the particle and the reference point. Two vectors can be added to make a third by pairwise adding their coordinates by ordinary arithmetic addition. Thus, if $\vec{A}$ and $\vec{P}$ are vectors, consisting of a row of numbers $(a, b, c, \cdots)$ and $(p, q, r, \cdots)$, then $\vec{A} + \vec{P}$ is also a vector, corresponding to the row $(a + p, b + q, c + r, \cdots)$.

About the 'when', people have speculated and philosophized for aeons. The necessity for a 'when' is not obvious, and in an imagined universe one may perhaps be able to do without it. In the actual Universe we find that measurements of the position vector are not unique without the specification of another quantity, called `time`. It turns out that time is a scalar: a single number, obeying the customary $2 + 2 = 4$, suffices to describe it.

3.6 The equation of motion

In order to find a quantitative answer to 'where is what when', we must find a prescription that tells us uniquely what position corresponds to what time. Let us take the difference between the position at an arbitrary time and the position a very small amount of time later (in Chapter 4 we will see what is meant by 'small', both in space and in time). We take this difference and divide it by the time elapsed between positions. Thus we obtain the rate of change of the position, called the `velocity`. By definition, (velocity)

= (position difference)/(time elapsed), where it is assumed that the elapsed time is taken to be arbitrarily small. Because the velocity is derived from a difference of two positions, it is a string of three numbers, a 'three-vector' like the position. We can visualize this vector as an arrow in space, pointing in the direction of the change in position. The length of the arrow – called speed – represents the amount of distance covered in one second along the direction indicated.

The velocities of particles may change over the course of time. In order to find a quantitative answer to 'where is what when', we must try to find out what velocity corresponds to what time. Exactly as above, we can define the rate of change of the velocity: the difference between the velocity at an arbitrary time and the velocity a small interval of time later, divided by the duration of that interval. This quantity, called acceleration, is obviously also a three-vector. By definition, (acceleration) = (velocity difference)/(time elapsed), again with an infinitesimally small time lapse.

By now, you may wonder if this game will ever stop, or if we will just mindlessly continue to define the rate of change of the acceleration, and so forth. We must somehow connect these vector beasts with things that happen in reality. In order to do this, consider the following arrangement (compare the Descartes slingshot in Fig. 1.1). On a perfectly flat and smooth horizontal table, we have a particle circling around at the end of a string which is so thin that its mass can be neglected. What is the motion of the particle (Fig. 3.16) if the string is suddenly cut?

A few possibilities are shown: the particle may continue in a circle, it may spiral outward, it may move outward in a straight line through the point where the string is attached, it may fly off on a straight tangent, it may stop dead in its track, or what not. It should be emphasized that it is not obvious what will happen; if you were to construct your own universe, any of these alternatives – or any other you take a fancy to – might occur. In the actual Universe we find, experimentally, that the released particle flies off on a straight tangent. This observed fact, and others similar to it, led Descartes to propose the law of inertia introduced earlier: an object moves in a straight line with constant speed unless a force acts on it. Using the definitions we encountered above, this can also be expressed as: *an object moves with constant velocity unless a force acts on it.*

We can formulate Descartes's law in terms of our three-vectors, by saying that the velocity vector (called $\vec{v}$) is a constant, unless a force $\vec{F}$ acts on it. Because we observe that a force is responsible for the *change* in velocity, we naturally seek a connection between the force and the rate of change of $\vec{v}$: there must be a relationship between $\vec{F}$ and the accel-

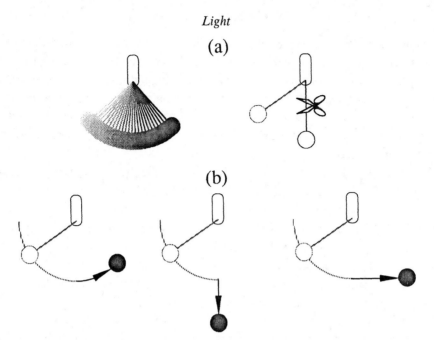

Fig. 3.16 Examples of possible outcomes of the slingshot experiment envisaged by Descartes: What happens to a pendulum (a) when its string is cut at exactly the point when it passes through the vertical? The panels (b) show three possible results. Which one is correct?

eration $\vec{a}$. We presume that $\vec{F}$ is proportional to $\vec{a}$, so that the force is also a three-vector. If we call the constant of proportionality m, we finally obtain $\vec{F} = m\vec{a}$. The constant m is the mass of the particle. The connection between force and acceleration is called the equation of motion; it tells us what acceleration we get if a force is applied to a particle.

Since the acceleration is the rate of change of $\vec{v}$, we can reconstruct by means of $\vec{a}$ how $\vec{v}$ changes in the course of time, if the velocity is given at an arbitrary initial instant. Once we know $\vec{v}$, which is the rate of change of the position, we can reconstruct how $\vec{x}$ changes over the course of time, if the position is given at the same initial time as the initial velocity. After these reconstruction steps, we have finally answered the question 'where is what when?', and we see that the answer is encoded in the equation of motion. The only ingredients we need are: the initial position; the initial velocity; the mass of the particle; plus a prescription for the force, and the equation of motion does the rest.

3.7 Momentum and energy

A careful study of the equation of motion shows that two quantities associated with a moving particle are particularly important. One is obtained by multiplying the velocity of the particle with its mass, which gives $m\vec{v}$. This is obviously a three-vector, and it is called `momentum`. The other is obtained by multiplying half the mass of the particle by the square of the particle's speed: $\frac{1}{2}mv^2$ (v is the length of the velocity vector $\vec{v}$). This is a scalar, called energy of motion or `kinetic energy`.

It is evident from the equation of motion that the momentum and the kinetic energy of a particle are changed by the action of a force. Therefore, we would expect that momentum and kinetic energy are not very useful quantities for keeping account of a particle's behaviour, unless we could somehow include the force in the accounting system. It turns out that the force can be taken into consideration, too, because a force is a mutual thing, acting between objects, rather than appearing magically from nowhere.

Suppose that we have a set of particles on which no force acts from the outside; in that case, the only forces that can occur are those among the particles themselves. Now calculate the total momentum of the system by adding up the momentum vectors of all the particles, and likewise find the net force on the system by adding all the force vectors among the particles. Because there is no interaction with the outside world, i.e. no net force, the sum of the mutual internal forces must be zero. When the force in the equation of motion is zero, the velocity does not change; nor does the momentum, so that *the total momentum is constant in time*. This fact is called the law of `momentum conservation`.

Next, we add up all the kinetic energies of the particles to obtain the total kinetic energy of the system. Because the kinetic energy is proportional to v^2, it depends on the speeds of the particles but not on the directions in which they move. Consequently, the kinetic energy is always positive, and we cannot use the vector addition trick used above to make the net force equal to zero: a row of positive numbers always adds up to something. Thus, the total kinetic energy is *not* in general a constant. However, we realize immediately that the speeds of the particles can only be changed by the combined action of the internal forces. Accordingly, for each particle in the system, we calculate how much its kinetic energy would change if we let it move by a unit distance (say one metre) under the influence of the forces exerted by all the other particles. Because this number indicates how much energy the particle could 'potentially' have gained, it is called `potential energy`. When the potential energy is included as an extra accounting unit, it can be shown that *the total energy is constant in time* if we define the total energy as the sum of the kinetic and potential energies. This fact is called the law of energy conservation.

The remarkable thing is that the total momentum and energy are constant no matter what the particles do to each other, as long as they are not exposed to outside forces. It can be shown that the above methods can be applied to systems that are not isolated from outside forces, provided that we take the potential energy of these forces into account, but we will not need that here.

These are the most famous of all conservation laws; we will encounter others later. The above discussion leans heavily on the classical-mechanics distinction between 'matter' and 'force', which leads us to 'kinetic' and 'potential' energy. When I first heard these terms in high school, I was extremely puzzled: why the dickens would there be *two* kinds of energy? If these are really two different things, why are we allowed to add them? Indeed, why *must* we add them to obtain a good conservation law? Why not *one* energy, considering the fact that kinetic and potential energy conspire to add up to a constant total? Then, more than a decade later, I learned that 'force fields' consist of quanta, so that the forces themselves can be considered as swarms of particles.†

If both matter and force are due to the behaviour of particles, they can be treated on an equal footing: roughly speaking, what is called $\vec{F}$ in the equation of motion refers to the virtual bosons in the system, and m refers to the fermions. I hope that you are as impressed as I am with the beautiful unity of this view. Gone are the two different forms of energy. In their stead, one finds but one quantity, called the action, that summarizes the behaviour of quanta. As we will see in Chapter 6, the action is intimately related to the phase of a quantum.

From the above conservation laws, and from the equations of motion, we can work out the details of collisions between particles. Now it is observed that, if we pass a radiation beam of waves with short wavelengths (say a bundle of X-rays) through a cloud of electrons, the electrons are knocked about in exactly the way we would expect if they were hit by a stream of particles. Apparently, the electrons recoil off bits of radiation precisely according to the laws of momentum and energy conservation in a two-particle collision. Therefore, we must conclude that a beam of light consists of a stream of *photons*, particles of light.

† This process is sometimes called 'second quantization', the 'first' quantization referring to the introduction of the quantum behaviour of the 'matter' particles.

4

Maybe I'm Heisenberg

4.1 Neither particle nor wave

The information about the particle behaviour and the wave behaviour of light, presented in the preceding chapter, forms an extremely peculiar combination when taken together. And yet both are true: the experiments have been done as described, with the outcome as stated. There are experiments that tell us clearly that light consists of waves; there are others that tell us, equally clearly, that light consists of particles. Somehow we must find a description of light that incorporates both kinds of behaviour.

This is a formidable undertaking, but there is one cheerful fact: in absolutely *none* of the experiments that have ever been done did the particle and the wave behaviours occur *simultaneously*. Because this is a fact that covers the entire history of physics, we feel justified in assuming that this is a general feature of Nature, and that there cannot ever be an experiment in which the wave and particle behaviours both occur at the same time. This is a generalization, for we can never expect to do all possible experiments. This generalizing hypothesis was first announced by Bohr, who had the wonderful combination of daring and insight to postulate that the observation of certain quantities excludes the simultaneous observation of certain others. A pair of such observables was called `complementary` by Bohr; in the current professional literature, one usually finds the expression 'conjugate variables' or 'conjugate operators'.

The hypothesis that wave and particle behaviours can never be observed in the same measurement is an expression of Bohr's principle of complementarity. The implication of this principle is encouraging but awesome: there need not be a conflict between the particle and the wave aspects of light, provided that we accept that Nature 'knows' whether we are observing the particle behaviour or the wave behaviour. The physical system under study can only 'know' what we are up to if *the observation necessarily influences what we observe.*

In one experimental setup, light shows its wave character, in another it shows its particle guise: it is the nature of the experiment that determines the outcome. We must drop the assumption that light can reveal, even in principle, all aspects of its 'self' – whatever that might be – simultaneously. This strikes a direct blow at the classical presumption that physical objects *have* a 'self' that is there, whether or not they interact with the outside world.

That seems odd, but it isn't really. All it means is that we drop the presumption that we can make statements about things we have not observed. Instead, we must adopt a certain principle of modesty, namely that we should refrain from pronouncements on the 'true nature' of a physical object, and that we should only speak about the outcome of actual observations. To many people, this principle of modesty seems entirely plausible, indeed obvious; but some – occasionally called 'realists' – wish to stick to the notion that there is such a thing as 'external reality', whether it is observed or not.

In realist's lingo, a phrase such as 'a photon exists only when someone observes it' appeals strongly to the prejudices we acquired by living in the coarse world of large-scale physics. But in less emotionally charged language, the quantum behaviour of Nature sounds much less odd. To begin with, the verb 'observe' does not necessarily imply the presence of a scientist in laboratory overalls. The word 'observe' is an anthropocentric expression, and we should replace it by 'interact'. Furthermore, 'someone' should be replaced by an expression for a particle or other physical system (which might, but need not, include an actual person). In that sense, an electron is 'someone'. Finally, the use of the verb 'exist' is totally unwarranted, and merely serves as a sly appeal to the biases instilled in us by the large-scale world. Bearing these points in mind, we can rephrase the above in a less emotional and more physical way: 'The presence of a quantum has physical consequences only when it participates in an interaction.' In other words: a non-participating quantum might as well not exist.†

Please remember that the words 'observing', 'disturbance', and all that, absolutely do not imply that quantum behaviour is only important in laboratory experiments. Properly speaking, one should use such sentences as 'an interaction between quantum systems' instead of 'an electron observed in a streamer chamber'. But such ponderous phrases, as heavy as those in philosophy, are ballast at best.

---

† However, 'participation' has some surprising aspects in the quantum world, in particular because quanta can be *indistinguishable*; I will discuss these aspects in Chapter 8.

Even so, the above is contrary to intuition in our everyday world. If, in spring, we put our skis in the attic until next winter, we can confidently talk about them in the summer: length, material, colour, weight; all these properties we believe to exist, down to the most minuscule detail, without having to climb up all those stairs to verify them. What is it, then, that makes quanta different?

4.2 Absolute size

The answer is simple: *size*. The key clause in the statement about the skis is *down to the most minuscule detail*. An experienced observer would raise an eyebrow at this: 'Oh yes? And how do you know? Have you ever really looked at your skis in infinitely small detail, when you got them out for the next winter season?' Surely we have restricted ourselves to inspecting the gross scratches made by rocks and sand, and have not plumbed the depths of despair about the damage at the atomic level; and it makes all the difference.

Let us recall Descartes, and his ideas about forces. He assumed that the Universe is filled, to the exclusion of all space, by the particles of some 'subtle matter'. Conceptually, this is surely possible, as long as we talk about the assembly of all particles; we might imagine, for example, that the distribution of particles would always look approximately the same, whatever the magnification, so that they fill space entirely with a kind of recursive surface (Fig. 4.1). In principle, particles could be self-similar: in that case, a particle would be built out of pieces that are similar to itself. But Nature does not appear to work that way: there is an absolute scale of smallness.

A self-similar nesting of particles cannot be a complete description, for in this picture an individual particle cannot exist. It must have zero extent, or else our description of Nature would not be complete: we would still have to face queries about the structure and stability of the particle in question. In the classical framework, we would have to assume that each particle is itself made up of smaller ones, and we would never be free of this plague of boxes-within-boxes unless we encountered an *absolute* scale of size below which no structure can be discernible. *If 'small' is only a relative concept, it is no good to explain big things in terms of the behaviour of smaller parts.*

What does size have to do with the statement that the observation necessarily influences what we observe? Consider some experimental setup. One would, guided by everyday experience, presume that the disturbance imparted to the observed object can be made arbitrarily small. But this can

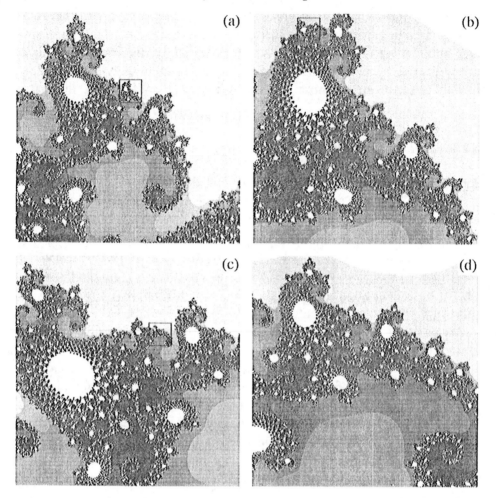

Fig. 4.1 A Julia set at various magnifications. A small square of each box is shown enlarged by a factor of ten in the next frame, so that the fourth box is 1000 times larger than the first. This can go on indefinitely, and still the picture will retain the same shape characteristics. Such a set is called self-similar. (Diagrams courtesy of the *Mandella* program for the Macintosh, by Jesse Jones.)

only be so if 'small' is a relative concept: we cannot make a disturbance arbitrarily small if there is an absolute scale of size. If the thing we are observing is such that, by being careful, the disturbance caused by the observation is made negligible, then the observed thing is 'big'. If the object we are observing is disturbed by the observation, no matter how careful we are, then it is 'small'. Thus an absolute scale of size exists if, in an experiment involving lengths smaller than that scale, the act of observation necessarily disturbs the observed. *The gentleness of any interaction has some lower bound*, which is a given limitation of our Universe that can never be circumvented by skill or care. This is the basis for the statement that 'taking a careful look' is impossible in the realm of the small.

Of course you will want to know *how* small. A metre? A millimetre, or even less? As we will see in Chapter 4, the length scale λ below which a particle with mass m can be called small is given by the De Broglie length $\lambda = 2\pi\hbar/mv$, where $\hbar$ is a universal number called Planck's constant, and v is the speed of the particle (actually, Planck used the symbol h, but at present we mostly use $\hbar = h/2\pi$). A grain of sugar dropping into your tea has a De Broglie length of roughly $\lambda = 10^{-28}$ metres, a thousand million million million million times smaller than the teacup. That is why I have been emphasizing how far out of our everyday experience the quantum world is!

If the De Broglie length of that sugar grain were the size of a marble, then you would be as large as the observable Universe. Therefore, you need not worry about the effects of smallness when you put sugar in a drink. However, a carbon atom in the sugar has $\lambda = 10^{-8}$ metres, which is only a hundred thousand times smaller than the physical size of the sugar grain. Thus you must expect that the effects of smallness become important when you want to describe the structure of sugar!

4.3 Through a glass, partly

A corollary to finding that the observation necessarily disturbs the observed is that there is a limit to the precision of observations. Consequently, *small means uncertain*. Let us consider the properties of this uncertainty. How can we picture a particle in such a way that its behaviour is of necessity a little bit uncertain? In the classical description given above we saw that the motion of a particle (under the influence of a given force) is exactly calculable if we know, at an arbitrary initial instant, the position and the velocity (or, more precisely, the momentum) of the particle. Therefore, we expect that its motion is *not* precisely calculable if there is some inescapable uncertainty about either the position, or the momentum, or both.

As soon as we admit vagueness in our description of Nature, it seems that we have hauled the Trojan Horse within the walls: uncertainty implies that any process can occur via several alternatives. If this uncertainty cannot be removed – and indeed it can't, if we want to make a meaningful distinction between big and small – then we must accept that *we will forever be unable to predict which alternative a process will take.*

Thus, we must accept `probability` as a basic quantity in our description of Nature, and a process cannot be described more precisely than by giving the sum of the probabilities of all its alternatives. As we will see throughout this story, the existence of alternatives, and especially the way in which these alternatives are added together, is one of the most essential aspects of the description of the realm of the very small.

If a small-size process can proceed via a (possibly infinite) number of alternatives, then we must accept that, due to the existence of an absolute scale of smallness, *the same causes do not always have the same effects*. Strange? Of course, but I'm sure that you are familiar with examples of this even in everyday life. When you stand in a brightly lit room in front of a window in the evening, you can see your own reflection in the glass. At the same time, someone standing outside can see you, too.

This is extremely odd: the light that is reflected towards your eyes leaves your face *under precisely the same conditions* as does the light that passes through the glass to the person outside! It is as if the light has a 'choice' about whether it can go through the glass or be reflected. The word 'choice' does not imply conscious action by the light, of course, but indicates that the interaction of the light with the windowpane takes place via several alternatives. These alternatives, when added together, produce effects collectively known as `interference`. For example, the 'choices' that a light ray has in reflection-and-transmission at surfaces (such as glass or water) can produce visible interference, as they do when producing the shimmering colours in a soap bubble.

As in the case of waves on water or in air, interference of quantum alternatives is calculated by adding these alternatives together. But how do we represent an alternative, and what do we mean by 'adding'? We have already seen many cases in which 'sum' does not mean simple arithmetic addition, for example in the vectorial addition of positions and velocities. Perhaps there exist a thousand ingenious ways in which you can add alternatives in your home-made universe, but which – if any – of these actually occurs must be determined by a shrewd guess based on observation.

Here I will introduce to you the method which physicists have discovered to describe quantum alternatives. The method was invented by a number of very smart physicists (in rough chronological order, the main ones are: Planck, Einstein, Bohr, De Broglie, Schrödinger, Dirac and Feynman) who tried to find a mathematical way to represent the baffling quantum behaviour of Nature. As you will see, it is not really very difficult at all, but it is phenomenally strange. And, in a very real sense, nobody – not even the people who invented this method – really understands it. That is to say, nobody knows whether the Universe *must* work in this way in order to work at all.

So don't worry if you do not understand the addition of quantum alternatives: nobody does, yet (but of course we will, some day). All we know is that the method we use has produced an immense quantity of results which have been verified with better accuracy than *any* others in science. Indeed, the *Standard Model* of particle physics, with which this book concludes, is the most accurate scientific theory to date.

Mathematically, a quantum alternative (i.e. a particular path of a quantum through space and time) is associated with a certain type of number called a *complex number*. Complex arithmetic is not very difficult, notwithstanding the name, but I wish to keep this book as free of algebra as possible. Therefore, I will resort to a trick which was first used by the French mathematician Argand.† Using Argand's geometrical ploy, a quantum alternative is drawn as an arrow, or two-dimensional vector. In physics parlance, this arrow is called an `amplitude` (by which, unfortunately, one doesn't mean precisely the same as the amplitude of a wave).

Such an arrow is two-dimensional, i.e. it has two components: a *length* and a *direction*. In the language of waves, these are associated with the *amplitude* and the *phase* of a wave. You will recall how a point, rotating on a circle, generates a wave; similarly, the length and the direction of the arrow conspire to produce the wave-like properties of quanta.

To each alternative, then, corresponds an amplitude, which is represented by an arrow. How do the alternatives add up? The amplitudes of all alternatives add by exactly the same kind of superposition that we observe in waves: two amplitudes *a* and *b* add as *a* + *b*. In terms of arrows, this means putting the arrows tip-to-tail, and then the sum is the arrow that connects the tail of the first arrow to the tip of the second (we will see more about all this later, e.g. in Fig. 6.3).

The experiment with the window in the evening showed us that the same causes can have different effects; therefore, quantum physics must take account of alternatives. The complete description of any process consists of the sum total of all its alternatives. Thus, any process can be represented by its 'total arrow', the total amplitude.

Finally, we must relate this amplitude to the events we actually observe; in the above example, one possibility is that we see our face, and another is that someone else outside sees our face, too. Thus, in one and the same

---

† Your library may not have his book *Essai sur une manière de représenter les quantités imaginaires dans les constructions géométriques*, considering that it was published in 1874. Too bad, because going back to the sources is marvellously instructive.

process, we get 'some of each': that is to say, there is a certain *probability* of each outcome occuring.

This probability can be derived from the arrow which represents the total amplitude of the process. It is obtained by taking the square of the length of the final arrow (which represents the sum total of all superposed amplitudes).

Say that a process has precisely two alternatives, with amplitudes a and b. The total process, in which all alternatives are taken into account, has amplitude $a + b$; the probability associated with the process is then $(a + b)^2$.

It is very important that the superposition should be *linear*, because linear superposition is the only way to add things on an exactly equal footing. When we make up a total of the form $a + b$, the addition is entirely 'democratic' and does not discriminate between a and b in any way. This must be so, because otherwise it would be possible to distinguish the alternatives corresponding to these amplitudes; and the whole point of the discussion of smallness is that alternatives are *not* distinguishable.

By the way, this gives the beginning of an answer to a question that you may have asked, on occasion: why is it that all electrons are the same, and all people (and the proverbial snowflakes) are different? It is because electrons are small and people are big. When something is small, it can no longer be distinguished from its siblings. This is why we often call such a small thing an 'elementary' particle.

It may seem a bit pedantic, but it is worth noting that I have not said that every small object 'is' a wave. All we say is that the qualification 'small' implies an inescapable uncertainty. Due to that uncertainty we must allow indistinguishable alternatives for the same process, and these alternatives have the same superposition properties as waves do. To repeat what was said when discussing water waves: when describing quanta, we can use the concepts and mathematics developed for classical waves, but that does not mean that quanta are ripples on an invisible cosmic pond.

In summary, the above discussion leads to the following remarkable outcome. In some interactions light behaves like a wave; in some, it behaves like a particle; these behaviours never occur simultaneously; thus, interactions have a 'choice' between alternatives; the existence of alternatives introduces uncertainty into the description of an interaction; this defines an absolute scale of smallness, in which 'small' means 'uncertain'; unavoidable uncertainty means the necessity of working with probabilities; in our Universe, these probabilities are obtained via a superposition process that is the same as that of waves. Thus, anything that is truly small has noticeable particle *and* wave aspects.

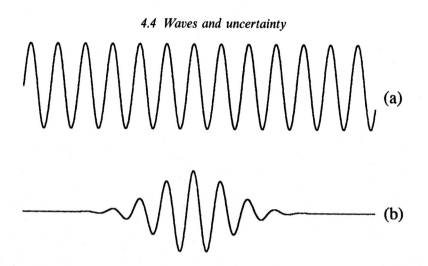

Fig. 4.2 An infinitely long wave (a) and a short wave packet (b). Each wave must be thought of as continuing indefinitely beyond the edges of the picture, but only the wave shown in (a) still has a finite amplitude at infinity.

However, it is not obvious that all these things can be fitted into a description that is free of contradictions. Reading again through the above list, we see that it can all be made to work – perhaps – if we can demonstrate, first, that there is a sense in which wave propagation is necessarily uncertain, and, second, that the quantities describing a wave can be uniquely linked with the position and the momentum of the particle it is supposed to describe. The latter is required because we have seen that, in classical mechanics, the position and the momentum of a particle are all we need to describe its motion under the influence of a given force.

4.4 Waves and uncertainty

Now what is it, in a wave, that is uncertain? The answer turns out to be: the combination of its position and its wavelength. If the wave is infinitely long (Fig. 4.2), we can determine its wavelength with absolute certainty, for example by the following procedure (Fig. 4.3). Count a large number of crests, say N; determine the distance D from the first crest to the last; then the wavelength is given by $\lambda = D/(N - 1)$. This is the way in which a gardener would calculate the average distance between N trees that stand in a row.

Because N can be made arbitrarily large in an infinite wave, the wavelength determination can be made arbitrarily precise. But what is the position of the wave? Because an infinite wave has no beginning, no middle and no

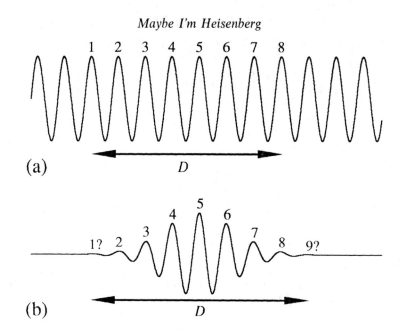

Fig. 4.3 Determining the wavelength by counting crests. Because the wave at (a) continues indefinitely with finite amplitude to either side, the counting distance D can be made arbitrarily large. This is not the case in (b), because that wave packet has a finite extent.

end, we cannot answer the question as to where the wave is. Hence, we know the wavelength exactly, but this exactitude was gained by forfeiting all knowledge of the whereabouts of the wave.

We may try to improve on this by picking a wave of finite extent, a wave packet. In that case, we have at least an approximate answer to the question as to where the wave is, because it certainly is not anywhere beyond either of the cutoffs of the packet. It is debatable where it is in between these points; we may arbitrarily choose the location of the highest point on the wave as the 'true' position, but it is not guaranteed that there is only one such point; and then, why not choose the lowest point? And in any case, the positions of all such special points shift as the wave propagates. So all we can say of a wave packet of length Δx is that its position is known to within the uncertainty Δx.

What of the wavelength? Again, we can count the number N of wave crests, measure the corresponding distance D, and conclude that $\lambda = D/(N -$ 1). But because the wave packet is finite, we cannot increase D and N arbitrarily; we always have that the distance D covered by all the crests in the packet is at most equal to the length Δx of the packet, and at the cutoff

points of the packet we cannot be sure that we aren't perhaps just missing one count in the determination of N. Therefore, the wavelength λ can in the best case be found to lie between the values $\Delta x/N$ and $\Delta x/(N-1)$. This means simply that $1/\lambda$ lies between $N/\Delta x$ and $(N-1)/\Delta x$. Thus, the uncertainty $\Delta(1/\lambda)$ in the value of the inverse wavelength, multiplied by the uncertainty Δx in the position of the wave, is at least unity: $\Delta x\,\Delta(1/\lambda) > 1$.†
The symbol '>' means 'larger than', whereas '<' stands for 'smaller than'.

The conclusion is, then, that the wavelength and the position of a wave *cannot be determined simultaneously with arbitrary precision*. A consideration similar to the one above shows that the same holds for the frequency of a wave and the time during which the wave passes by a fixed observer, such that the uncertainty Δf of the frequency and the uncertainty Δt of the time during which the wave passes by obey $\Delta f\,\Delta t > 1/2\pi$. These uncertainty relations hold for *all* waves; you are strongly encouraged to experiment with everyday waves, and with wave packets drawn on paper, to get a feeling for this.

4.5 The uncertainty relations

The final piece in the puzzle of particles and waves concerns the following: what mechanical properties of the particles of light (photons) are associated with the wavelength and the frequency of light waves? We cannot expect a philosophical answer to this question, but must turn to experiments to provide it. The crucial discovery was made by Einstein, who noted (in his discussion of the photoelectric effect) that the experimental results require that the two fundamental mechanical quantities, energy E and momentum p, are related to the wave frequency f and the wavelength λ by $E = 2\pi\hbar f$ and $p = 2\pi\hbar/\lambda$. The quantity $\hbar$ is Planck's constant.

If we combine these prescriptions with the above uncertainty properties of waves, we finally find that $\Delta x\,\Delta p > \hbar$ and that $\Delta E\,\Delta t > \hbar$. These equations are called Heisenberg's uncertainty relations. In words: the uncertainty of the position of any object, multiplied by the uncertainty of its momentum, gives a number which is bigger than Planck's constant. Similarly for the product of the uncertainty of energy and time: the uncertainty of the energy of any object, multiplied by the uncertainty of the time during which it resides at a given point, gives a number which is bigger than Planck's constant.

How are these relations connected with our assertion that small means

---

† Actually, a precise calculation shows that the correct expression is a factor 2π larger, namely $\Delta x\,\Delta(2\pi/\lambda) > 1$, but our estimate is pretty close.

uncertain, or that an observation necessarily influences what we observe? Let us consider $\Delta x \, \Delta p > \hbar$. This expression says that if we locate the position of a quantum with an accuracy of Δx, a simultaneous determination of its momentum cannot give a precision better than $\Delta p = \hbar / \Delta x$. Thus, *small means uncertain*: the smaller the distance Δx within which the quantum is confined, the larger the uncertainty in the state of motion (i.e. the momentum) of the quantum.

Before proceeding, it should be noted that the uncertainty relations are statements about the *joint precision of two quantities*. A pair of quantities that are linked by an uncertainty relation are `complementary` in the sense defined by Bohr. Because it is only the *joint* precision of conjugates that is restricted, any *single* quantity can be detemined with arbitrary precision; but the greater the precision of one variable, the greater is the uncertainty of its conjugate partner. Later, we will see how this complementarity is responsible for the existence of stable atoms with finite size, even though they are built of nuclei and electrons that always attract each other (remember the collapse of Newton's gravitating universe!).

Figures 4.2 and 4.3 are illustrations; the actual behaviour of participants in uncertainty relations is more subtle. Due to relativity, we cannot see space and time as separate, nor can we make a strict distinction between momentum and energy. It then turns out that ΔE and Δt must be interpreted as the energy spread and the lifetime of a particle. I will use them in this sense below. In quantum parlance, particle properties such as position and momentum that occur in an uncertainty relation are called `conjugate variables` or `conjugate operators`.

4.6 De Broglie rules the waves

In the foregoing we have seen that in some interactions photons behave like waves, in some others they behave like particles, and they never do both simultaneously. We have assumed that the latter is a general feature of Nature, i.e. not only haven't we been able to produce an interaction in which a photon shows its particle and wave character at the same time, but indeed it is fundamentally impossible to do so. Only with this assumption can we escape the horns of the wave–particle dilemma.

Now we know from observations that photons can scatter off other particles, for example electrons. This implies that there can be a definite region in space and time when a photon and an electron are closely coupled; during the time interval Δt of the interaction, the positional uncertainty Δx of the photon is the same as that of the electron. Imagine that we could measure the momentum p and the energy E of the electron, before and after the collision, with arbitrary accuracy. Then we could also, via the conservation of momentum and energy, determine the p and the E of the photon precisely. Since the uncertainties Δx and Δt of the interaction between a photon and an electron in space and in time are not infinitely large, this alleged precise determination of momentum and energy would violate the uncertainty relations. In that case, it would be possible to do an experiment that shows the particle and the wave behaviours of light simultaneously. This would lead to insurmountable paradoxes, so we are obliged to save the uncertainty relations for photons by requiring that *Heisenberg uncertainty is universal*, and holds (for example) for electrons as well.

In this way, the uncertainty principle spreads, like an epidemic, by contagion: it is observed to hold for photons; thus, it must apply to particles such as electrons and protons which can interact with photons; since neutrons interact with protons, the former obey the same uncertainty relations as the latter; and so forth. If the uncertainty relations are universal, we cannot determine exactly what happens to an electron in a particular collision, and the scattering will proceed with uncertainties Δp and ΔE as well as Δx and Δt.

Thus we reach the astonishing conclusion that we can only keep out of trouble if we extend the duality of particle and wave behaviour from photons to all of Nature. Accordingly, there must be experiments that show electrons to behave like particles, and there must be some that show them to behave like waves! All that remains to be done to complete the picture is to find the connection between mechanical aspects (momentum p and energy E) and wave aspects (wavelength λ and frequency f). It was first pointed out by De Broglie† that it is simplest to use the prescription‡ that we know to be correct for photons, namely $p = 2\pi\hbar/\lambda$ and $E = 2\pi\hbar f$. Note that this is a hypothesis, albeit a very plausible one; only experiments can decide if it is wrong and, as it happens, experiments have vindicated De Broglie's assumption.

Let us see what uncertainty means in the case of our skis, mentioned above. The size of a ski is between 10 cm and 2 m, say 1 m on average. Thus, the uncertainty in the position is of the order of $\Delta x = 1$ m. Now

---

† This famous name is probably one of the most mispronounced names of all time. It derives from the older Italian family name of Broglia, and is pronounced in a vaguely Italian way: *De Broy-eh*, with the stress on the *oy*. But never mind – whoever did the work, the only name people remember is Einstein, which fortunately is easy to pronounce.

‡ Some readers may have noticed that I have been writing the momentum p as if it were a scalar, whereas it is really a three-vector. This was done just to simplify the argument. Actually, the momentum $\vec{p}$ is related to the `wave vector` $\vec{k}$; this vector points in the direction in which the wave propagates, and its length is equal to $2\pi/\lambda$. Thus, $\vec{p} = \hbar\vec{k}$ and $\Delta\vec{x}\,\Delta\vec{p} > \hbar$.

Planck's constant $\hbar = 1.055 \times 10^{-34}$ kg m^2/s, so that the uncertainty in momentum is approximately 10^{-34} kg m/s. The mass of a ski is of the order of a kilogram, and because momentum equals mass times velocity we find that the uncertainty of the speed of a ski in our attic is 10^{-34} m/s. At that rate, it wouldn't get very far: if it had been stored during the entire lifetime of the Universe, it would not have moved spontaneously more than a millionth of an atom's thickness!

But now let us see what it means in the case of a single atom, say an atom of carbon in the vinyl sole of the ski. Its size is about a billionth of a metre: $\Delta x = 10^{-9}$ m. Therefore $\Delta p = 10^{-25}$ kg m/s; the atomic mass is approximately 10^{-26} kg, so that the uncertainty of the speed of the atom is 10 m/s. Of course, that speed would not be maintained over a whole second, but would change in an indeterminate way. Who knows, then, where that one atom has been while you were relating your skiing adventures to your skeptical friends?

Indeed, small means uncertain. Because I'm big, the uncertainty of my whereabouts is negligible, and the probability that I'm Heisenberg does not differ appreciably from zero. Because photons are small, uncertainty about their behaviour is rife. In our discussion of mechanics, we saw that the equation of motion of a particle allows us to predict its future behaviour exactly, provided that its initial position and momentum were given at an arbitrary point in time. Now Heisenberg's first uncertainty relation says that these two quantities *cannot* be determined simultaneously with arbitrary precision. Therefore, the motion of a particle cannot be predicted exactly. We can do a very good job if the particle is big, but must do very badly if it is small.

4.7 Quanta

From now on we can treat photons, electrons and all that stuff on an equal footing, so we will just use the word quantum or 'particle' as the generic name of such objects. *All pictures of quanta are necessarily metaphorical*; because of the uncertainty relations, we cannot make a proper drawing of a quantum, unless we were to draw it very, very small, and then the drawing would be a quantum itself. Because there is an absolute scale of smallness, making proper scale models or enlargements of quanta is out of the question! If we are careful to keep that limitation in mind, we can imagine a quantum (in one spatial dimension) as a wave packet.

In Fig. 4.3, the particle aspect of the quantum is provided by the envelope of the curve, which shows the extent in space of the packet; the wave aspect is associated with the oscillations, showing the wavelength. It is clear from the diagram that these two aspects are inseparable; just try for yourself to draw a wave without an envelope, or vice versa. You will soon notice that in the first case you are obliged to continue the wave indefinitely at both ends, so that you have a wave on your hands without a definite location in space (even though it is very clear what the wavelength is). In the second case you know very well where the wave is, but it has no definite wavelength. Such a picture is very dangerous: we must never forget that it is merely a symbolic summary of the duality of wave and particle behaviour.

One major reason that a picture of a quantum can never be more than symbolic, is that the wave amplitude does not behave like a real number such as 2.718, $\sqrt{2}$, π and so forth. Instead, the amplitude is determined by a set of *two* numbers. If we insist, these can be kept separate in our calculations, but that would entail a lot of superfluous bookkeeping. It turns out to be easier to lump the two amplitude numbers together to form what mathematicians call a `complex number` (in olden days occasionally referred to as an 'imaginary' number). The reason that this works is implicit in the De Broglie conditions, as was first shown by Schrödinger. We won't have to worry about the complex-number behaviour of quanta in our discussions.

With the formulation of the De Broglie conditions as universal relationships, we have finally reached a point where we can say precisely what we mean by 'small', because we now have a measure of size associated with any particle of mass m. This size is, of course, the De Broglie wavelength $\lambda = 2\pi\hbar/mv$ (remember that the momentum of a particle is the product $m\vec{v}$ of its mass and its velocity, or mv in one dimension). Given that the universal constant $\hbar = 1.055 \times 10^{-34}$ kg m^2/s, it may amuse you to calculate when objects of a given mass and speed can be called small. Just punch the mass m (in kilograms) and the speed v (in metres per second) into your calculator. We have already seen that the De Broglie wavelength of a freely falling grain of sugar is astonishingly small, about ten trillion trillion times smaller than the grain itself. But an electron moving at one metre per second has a wavelength of a few millimetres, which is quite a respectable size for a subatomic beast!

The De Broglie conditions and the Heisenberg uncertainty relations forge a connection between particle and wave behaviours, and show to what extent small means uncertain. But they do not show how a free quantum moves. Is there anything that can take the place of the mechanical equation of motion?

4.8 Where is a quantum when?

The theory that describes the motion of small objects is called `quantum mechanics`, as opposed to 'classical' mechanics, which is valid only in the realm of the large. It can be proved that the laws of quantum mechanics gradually and smoothly become identical with† the laws of classical mechanics if we consider larger and larger length scales.

As in the case of classical mechanics, we cannot talk about forces in quantum mechanics until we have decided what an object does if forces are absent. The first order of business is to find out what quantum path corresponds to the classical motion with constant velocity. By the way, many contemporary attempts at guessing the detailed behaviour of particles and forces first consider free particles only. It may seem bizarre to write papers on non-interacting particles, but this approach has a venerable history, going all the way back to Galileo and Descartes.

Remembering the question 'where is what when', we see that the rephrasing 'where is a quantum when' forces us to define what 'where' is supposed to mean. As far as is known today, the 'when' doesn't require any revision, but because of the particle–wave duality of the quantum we must make sure that our prescription for 'where' is valid for particles as well as for waves.

At first, this seems an insuperable problem: looking at the wave packet picture, we are at a loss. If we only consider the particle aspect (i.e. the envelope of the packet), the answer seems clear: the position of the particle is at the place where the envelope reaches a maximum. However, there is really nothing to prevent us from constructing wave packets that have envelopes with two peaks; and what then? Moreover, the presence of the wave under the envelope suggests that the particle is somehow present in the whole uncertainty range between the cutoffs at the ends of the packet. If we only consider the wave aspect, the answer seems clear also: the position of the wave is at the place where it reaches a maximum. But objections similar to those above can be raised. Moreover, a careful study of the relationship between the envelope and the wave shows that the wave crests do not move at the same speed as the envelope, so that we cannot use both of the above definitions of motion simultaneously, even if we could make them unambiguous. By the way, the different propagation of the wiggles and

---

† There are some interesting exceptions: some quantum rules, such as the Pauli principle (which we discussed when talking about matter and force, and which we will encounter again later), do not depend on a length scale because they do not refer to a physical dimension but to a purely algebraical or geometrical property. The quantum effects due to such rules do not disappear as the length scale becomes larger. For example, the apparent solidity of matter and the existence of chemically different elements are due to the Pauli rule; these are large-scale quantum effects that have no classical analogue.

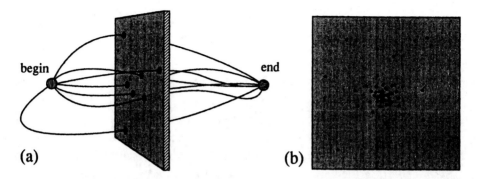

Fig. 4.4 Woolly path of a quantum. (a) Shown as a sheaf of Feynman paths through space. (b) Shown as the outcome of many position determinations on some plane between the initial and final positions. Each Feynman path is a legitimate alternative, even if it goes to Sirius and back (not shown here). In actual calculations, a Feynman path is a trace through space-time.

their envelope is present in most waves, and is not restricted to quanta; this effect is called `dispersion`, and you can observe it yourself in the case of surface waves on water.

We cannot feel justified in designating a single point on the wave packet as 'the' position of the quantum. Indeed, if we were to single out such a point, we would offend against the uncertainty relations. But although we cannot indicate a single point, we *can* designate a *range* of points within which the particle will surely be found; and it is clear that the height of the envelope at a given point is somehow related to the 'amount of presence' of a particle at that point. Because it cannot be indicated by a single location, we are led to a prescription for 'where' that acknowledges the fact that the quantum is, in a sense, distributed in space: the `probability` that the quantum is at a given point is proportional to the square of the amplitude† of the wave packet envelope at that point. We must use the square because otherwise we would have to contend with negative probabilities.

Thus, because the 'where' must be interpreted as a probability distribution, the path of a quantum (Fig. 4.4) cannot be idealized to an infinitely thin line, but may be thought of (metaphorically) as being a woolly one, indicating the distribution in space of the chance of finding the quantum somewhere. In fact, the concept of a 'woolly path' is fundamental to the quantum behaviour of Nature. Every two points in space are connected by a possible

† Because the wave amplitude is a complex number, we must be careful to define 'square' correctly. To be precise, we should use 'the sum of the squares of the real and the imaginary part' of the number, but we will not need so much detail in the discussion.

path; associated with each path is a phase. We will encounter these Feynman paths in Chapter 6.

It is found that the classical path of constant velocity, first identified by Descartes as the trajectory of a particle on which no force acts, corresponds to the path of largest probability in quantum motion. In fact, under most circumstances the woolliness of the path is imperceptible to us in our gross world. An electron that is catapulted out of a glowing metal wire in your laboratory is not very likely to be found on the Moon. But when you are measuring the electron's location you cannot be sure, in advance of the measurement, that it is *not* 330 000 kilometres away from Earth.

By interpreting the wave packet as the source of a probability distribution, we find that it provides information about the probability that *one* quantum is at a given point, and *not* the number of quanta we expect to find at that point. Moreover, the probability distribution can tell us nothing about 'the' path of a quantum, but that is no surprise: 'path' is a concept that refers strictly to large particles, and is useless here. Only the ghostly vestige of a path is found in the interference patterns.

5

Catch a falling quantum

✦

5.1 An experiment with electrons

The preceding chapters described the quantum behaviour of Nature in a general way. In order to show quantum uncertainty at work, it is useful to look at an actual experiment in detail. This demonstrates the practical consequences of the particle–wave duality. To keep things uncluttered, the experiment is only an idealized example, but it will serve to show when quantum interference occurs (wave behaviour) and when it does not (particle behaviour). I will try to demonstrate the intimate connection between the properties of waves and linear superposition, and the consequent uncertainty in the motion of quanta.

For once I will try to be precise, so this exposition will be rather more technical than elsewhere. Those who wish to do so, may skip to Chapter 6 without much damage to their education. Before we begin, let me repeat the warning about the generality of analyses such as the one to be given here: I speak about 'experiments' or 'observations' only in order to make a connection with what a real person could do in a real laboratory, and not because the particle–wave duality is only a laboratory phenomenon that requires a conscious observer. Even though an electron does not wear a lab coat, it is subject to the uncertainty relations when it 'observes' other quanta.

Let us construct a gun (Fig. 5.1) that fires off electrons of fixed energy, one at a time, from a very small region in space. Some distance beyond this source we place a flat barrier that is a perfect absorber of electrons. In this screen we drill two very small holes. Some distance behind the barrier we place a row of detectors which are so sensitive that the passage of an electron through one will surely be noticed.

Now we close one hole (call it number 2) with an electron-absorbing plug, leave hole number 1 open, and fire the gun. We observe that, after a fraction

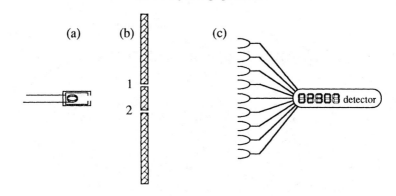

Fig. 5.1 Two-hole interference setup. Shown at (a) is an electron gun. The screen at (b) is opaque except for two very small holes. The particles are detected at (c) by a row of electron detectors.

of a second, each firing of the gun has one of two possible consequences: either nothing happens, or one detector signals the arrival of one whole electron. There is never a partial response of a detector, or a simultaneous response by two detectors (assuming the equipment works properly). Because of all this, we say that the electrons behave as particles; our mental image of this stage of the experiment is that the electron, shot out of the gun, either hits the screen and is absorbed, or hits the hole and is scattered from there into one of the detectors. After the gun has been fired a large number of times, we count the total number of registrations at each detector, and obtain a distribution as shown in Fig. 5.2. We can make our detectors very small, so we are justified in drawing a smooth line through the data points to represent the arrival probability of electrons at various positions. I will call this curve P_1.

If we were to interpret this part of the experiment by using the wave aspect of the electron, we would say that when the gun fires, it launches a wave packet; because the packet originates in such a small region (small compared with the wavelength of the packet, that is), it propagates as a spherical wave, by virtue of Huygens's principle. Part of the wave front hits the hole, and because the hole is small, the transmitted wave is also spherical, centred on the hole, by Huygens's construction. Finally, the wave hits the detectors; the probability prescription says that the square of the amplitude gives the probability of detecting the electron, and we find the same curve P_1 as above. Note that we observe that there is never a partial response of any detector or a multiple response of several detectors. Therefore, in order not to come into conflict with the particle aspect, a detection means that

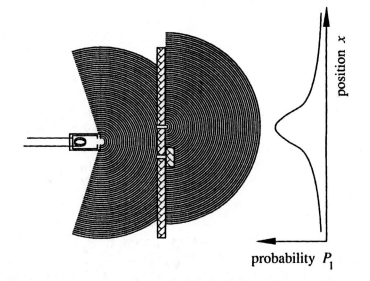

position x

probability P_1

Fig. 5.2 Graph of P_1, which is the particle distribution observed with hole 1 open. The height of the curve at point x on the axis indicates the probability P_1 that, at the location x in space, an electron is detected.

the electron cannot be thought of as being all over the place, but must be entirely in one detector.

5.2 Indeterminacy

This phenomenon, which is sometimes called the 'collapse of the wave packet', is due to the unavoidable disturbance of the electron by the observation. Notice, incidentally, that the wave packet also 'collapses' at the hole in the screen.

The expression 'collapse' of the wave packet is a frightful misnomer. There is nothing that collapses in the sense that the electron 'is' distributed in one particular way at some moment, and 'is' otherwise distributed an instant later. It cannot be overemphasized that the wave packet that describes a quantum says absolutely nothing at all about 'where' the quantum 'is', any more than a deck of shuffled cards says where the Queen of Hearts 'is' (no peeking!). Rather, the wave packet contains a summary of the probability distribution of finding the electron somewhere in space and time. Thus, a single detection of an electron is not an offspring of the wave packet at some earlier time, but is merely one instance of the whole distribution of possibilities described by the wave packet.

The detection is a single draw from an infinity of possibilities. If I draw a card at random from a deck, and it turns out to be the Queen of Hearts, I do not say that the deck has somehow 'collapsed to become the Queen of Hearts'. A wave packet is analogous to a recipe for manufacturing a deck of cards; an observation corresponds to a random draw from that deck.

The randomness is not 'caused' by anything (such as clumsiness on the part of the experimenter), but exists solely because in our Universe there is an absolute scale of smallness, wherefore observations necessarily perturb the observed. The outward manifestation of the resulting uncertainty is the randomness of the draw. At which detector the wave packet 'collapses' is indeterminate; all we know is that many repetitions of the experiment yield the distribution of detections as shown. Therefore, we would do best to interpret the ugly term 'collapse of the wave packet' as a metaphor for the truism 'hindsight has perfect eyes'. *After* a detection, it is no great achievement to solemnly declare what the electron 'did', but *before* the fact, all we can do is construct the probability distribution of many such observations.

The word indeterminate in the above is used deliberately to distinguish our state of quantum knowledge from the one described by *unknown*. The uncertainty relations show that the position and the momentum of a quantum cannot both be measured exactly at the same time. This makes the future motion of the quantum unavoidably indeterminate, which is quite distinct from unknown: the latter would imply that we have been careless, lazy, unlucky or stupid in our experiment.

5.3 Interference

Now we close hole number 1, open hole number 2, and repeat the experiment. We obtain the same probability distribution (Fig. 5.3), but shifted by the distance between the holes; the shifted distribution will be called P_2. This is entirely what we expect, and this part of the experiment teaches us nothing new. But what if we were to leave both holes open? On the basis of our everyday experience with probabilities, we expect that the resulting distribution will be $P_{12} = P_1 + P_2$. But when we perform the experiment, we get something wildly different (Fig. 5.4): the curve P, which bobs up and down while tapering off into the distance.

This demonstrates that quantum probabilities do not add like ordinary single numbers. That is nothing to be dismayed by; we have already seen that many everyday physical quantities, such as positions and velocities, do not add like single numbers either. The experiment tells us that the various

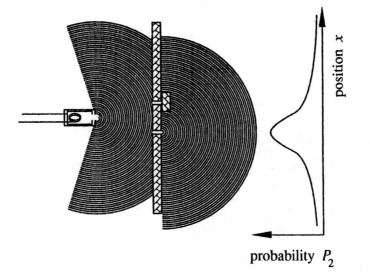

position x

probability P_2

Fig. 5.3 Graph of P_2, the particle distribution with hole 2 open.

parts of a quantum wave can show interference; this is an observational reflection of the fact that the addition of probabilities is more complicated than adding single numbers.

The graph of the distribution P shown in Fig. 5.4 may seem a little abstract, so let us see what would come out in an actual experiment. When the detectors register the impacts of the individual particles, we do not see much structure at first (Fig. 5.5); in fact, a few hundred impacts give a distribution that looks rather like P_{12}. But when more impacts accumulate, it becomes clear that there are narrow zones where there are almost no impacts at all, separated by ridges where many particles are detected.

Let us say that a_1 is the height of the wave packet envelope in the first part of the experiment, a_2 in the second, and a in the final part where both holes are open. More precisely: a_1 is the amplitude of the first alternative (passage through hole 1), a_2 is the amplitude of the second alternative, and a is the amplitude of the whole process, in which both alternatives are present. According to the probability interpretation of the wave packet, the likelihood P_1 that the electron will be at a given location is equal to the square of the amplitude: $P_1 = a_1^2$. Similarly, we have $P_2 = a_2^2$. Because of linear superposition, we now get $a = a_1 + a_2$, so that $P = (a_1 + a_2)^2 = a_1^2 + 2a_1a_2 + a_2^2$; and this is *very* different from $P_{12} = P_1 + P_2 = a_1^2 + a_2^2$! We do not get

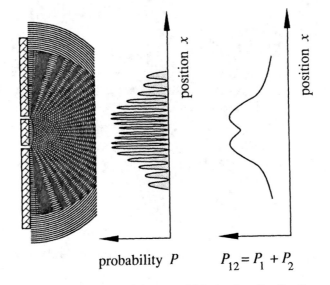

probability *P* $P_{12} = P_1 + P_2$

Fig. 5.4 Rightmost curve: graph of P_{12}, which is the distribution expected from summing P_1 and P_2. The curve on the left shows the probability distribution P, which is the distribution actually observed when both holes are left open.

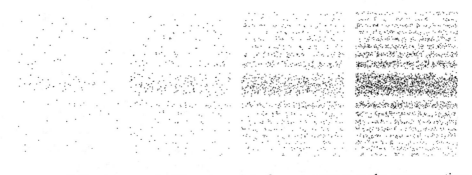

Fig. 5.5 Individual impacts build up to an interference pattern as the exposure time increases. This plot shows the interference pattern due to a single wide horizontal slit (which may be thought of as a row of adjacent narrow openings). We see the particle behaviour (left) because we observe individual impacts. We also see the wave behaviour, in the interference maxima and minima ('fringes') which become evident when we collect a great many impacts (right).

$P = P_{12}$ because of an `interference` term equal to the product $2a_1a_2$ that occurs in P but not in P_{12}.

Notice, by the way, that a_1 and a_2 are first added and then squared. Thus, P can be *zero*: under the proper circumstances, interference can be totally destructive, in which case there would be detectors in our experiment that never detect an electron at all, even though such a detector is in plain view of both holes! We will return to this in the next chapter.

Because we observe that the probability distribution with both holes open is not equal to the direct sum of the distributions in which each hole is open in turn, we must conclude that, when both holes are open, it is *wrong* to presume that the electron goes either through one hole or through the other. After all, 'going through' a precisely specified location is purely a particle concept, and, as the picture of the diffraction process shows, the interference works only if the wave propagates through *both* holes. Because of the complementarity inherent in waves, there are no experiments that show the particle and wave behaviours simultaneously, and this – as we have seen – led to Heisenberg's uncertainty relations. Consequently, it is false to think that it is *unknown* through which hole the quantum goes; it is indeterminate.

Any attempt to remove the indeterminacy must also remove the interference: an experiment that tries to ascertain through which hole the electron went is perfectly feasible, but any such experiment kills the interference. If we have an experiment in which both holes are certainly open, it is indeterminate whether the electron went through one hole or the other, and we observe interference. If we have an experiment in which only one hole is certainly open, but we forgot to write in the laboratory log book which one, then it is merely unknown which one it was, and we do not get interference. The whole thing hinges on the fact that *an observation of the actual path of a quantum excludes all alternatives except one*. If there is only one alternative, there obviously can be no interference between several alternatives.

The above can be summarized as: 'Want a definite path for the electron? Then no interference'. 'Want interference? Then you cannot know which path the electron took'. As was argued in the preceding chapters, this mutual exclusion saves us from contradictions, and the whole quantum construction hangs together without logical flaws. But it would be nice to demonstrate explicitly that any attempt to determine the electron's path destroys the interference pattern; this we will presently do.

5.4 Where does the electron go?

Let us observe the electron when it emerges from the screen with the holes. Observation means interaction, so we must provide a source of quanta to intercept the electron. It doesn't matter what kind of quanta we use, because De Broglie's and Heisenberg's relations hold universally. For definiteness, let

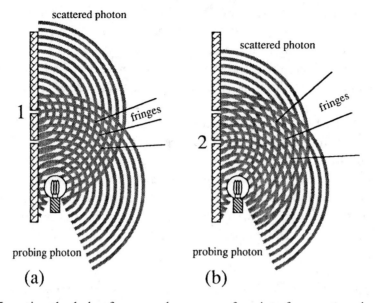

scattered photon

1

fringes

probing photon

(a)

scattered photon

2

fringes

probing photon

(b)

Fig. 5.6 Locating the hole of passage by means of an interference experiment. At (a), we see how a probing light wave interferes with itself when it is scattered behind hole 1. In this case, the fringes are closely spaced. At (b), we obtain a *different* interference pattern if the probing wave is scattered behind hole 2: the fringes are spaced more widely.

us say we use photons. We send a stream of photons along the screen, and lo! we observe that for every electron that arrives at the row of detectors, there is a photon–electron scattering event either behind hole 1 or behind hole 2, never behind both at once.

We can readily locate the hole at which a photon is scattered by performing an additional interference experiment (Fig. 5.6). If the light comes from a point source, it sends forth a spherical wave; upon scattering, we obtain another spherical wave, centred on the hole behind which the electron and the photon met each other. The scattered wave interferes with the direct wave from the light source. The interference pattern we observe when the scattering takes place at hole 1 is different from the one we observe when the electron and the photon scatter at hole 2. In this way, the experiment demonstrates without doubt that the electron does, in fact, pass either through hole 1 or through hole 2.

Because observation means interaction, the photon–electron scattering that reveals the location of the electron causes both quanta to recoil; the consequent uncertainty Δp of the momentum of the electron is roughly equal

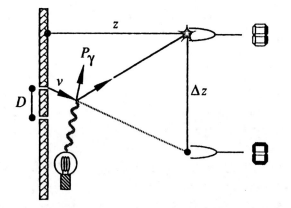

Fig. 5.7 Calculation of what happens when a probing photon knocks into the flying electron. The distance between the holes in the wall is D. The distance between the wall and the detectors is z. The electron comes out of a hole with velocity v and is knocked sideways over a distance Δz by a photon with momentum P_γ.

to the momentum p_γ of the photon: $p_\gamma = \Delta p$. Furthermore, we recall the De Broglie rule which connects the momentum and the wavelength of a quantum by $p_\gamma = 2\pi\hbar/\lambda$. If the momentum of the electron before the scattering is p, we conclude that the relative perturbation due to the measurement is $\Delta p/p = 2\pi\hbar/p\lambda$. By definition, momentum equals mass times velocity, so we find $\Delta p/p = m\,\Delta v/mv = \Delta v/v$. If we let the electron fly away after the scattering during a time t, it covers a distance $z = tv$; thus, the scattering knocks it sideways over a distance $\Delta z = t\,\Delta v = z\,\Delta v/v$. Accordingly, if the detectors are placed at a distance z behind the holes (Fig. 5.7), the observation of the electron causes an uncertainty Δz in the position at which the electron arrives at the detectors, and $\Delta z = z\,\Delta v/v = z\,\Delta p/p = 2\pi z\hbar/p\lambda$.

Is this a really bad knock? That depends on the spacing between the interference fringes. In the chapter on waves and interference, we saw in detail that the spacing of the fringes becomes smaller as the spacing between the holes becomes bigger, and vice versa (if you have made a transparent photocopy of the spherical wave, some shifting of it with respect to its original will make this dramatically evident). Thus, if the spacing between the holes is small compared with the wavelength, no interference experiment can distinguish between them. Indeed, this is the same as saying – following Huygens – that scattering from a very small region produces a spherical wave. Let us say that D is the spacing between the holes, z is the distance to the detectors, and λ_e is the wavelength of the interfering electron wave. Then a careful look at the interference patterns (see Fig. 3.13) reveals that the distance A between the fringes obeys the relationship $A/z = \lambda_e/D$ or, equivalently, $A = z\lambda_e/D$. By using the De Broglie rule $p = 2\pi\hbar/\lambda_e$ for the electron, this can also be written as $A = 2\pi z\hbar/pD$.

So now we have two numbers: the distance Δz over which the electron is knocked sideways by the observation, and the spacing A between the electron interference fringes in the undisturbed experiment. If Δz is much smaller than A, we are safe, because then the knock due to the observation is too gentle to obliterate the interference pattern. Putting together the expressions for Δz and for A, we require that $2\pi z\hbar/p\lambda$ must be smaller than $2\pi z\hbar/pD$; this is achieved when $1/\lambda$ is less than $1/D$. Therefore, we must require that D is less than λ.

So far, then, the conclusion is this: it is easy to observe through which hole the electron passes, but we must make sure that in the observation we use photons with a wavelength λ that is *bigger* than the spacing D between the holes. Otherwise, the interference pattern will be wiped out. There is absolutely nothing to prevent us from using photons with a wavelength as long as we please, which means that we can make the disturbance as small as desired. Does that also mean that we can keep the interference pattern intact when we observe through which hole the electron goes?

No! In order to be able to do this, we must be able to determine behind which hole the probing photon was scattered; in other words, we must be able to distinguish the scattering pattern of the photon behind hole 1 from the pattern we obtain when the photon is scattered behind hole 2. These two patterns are easily distinguishable, but only as long as the photon wavelength λ is considerably *less* than the hole spacing D! So either we have that λ is smaller than D, in which case we can determine whether the scattered photon comes from hole 1 or 2, and we wipe out the interference fringes; or else we choose λ to be larger than D, in which case the interference fringes remain visible, and we cannot discern behind which hole the photon was scattered. This is complementarity in action.

In the above, we have analysed one specific experiment, but in the analysis we have only used some general properties that *all* waves (even ordinary sound or water waves) share, plus De Broglie's connection between mechanical quantities (momentum and energy) and wave quantities (wavelength and frequency). Consequently, the results obtained are also general, and we can state Heisenberg's uncertainty principle in an appropriately general form: *if a process can follow more than one alternative, then any determination of which alternative is actually followed destroys the interference between the alternatives*, because this determination necessarily reduces the number of actual alternatives to one.

The experiment we have analysed is a special case of this rule, and can be summarized thus: any determination that a definite hole in the screen was hit by the electron, and that the other was certainly not hit by it, eliminates the influence of all but one point of the screen on the diffraction pattern behind it. This pattern must then be the same as if only the hole that was hit were present, i.e. the pattern is a spherical wave.

Some experimenting with the interference between a transparency of a spherical wave and its original shows that the foregoing conclusions are unavoidable because the closer we move the centres of the interfering spherical waves together, the more broadly does the interference pattern spread out. In fact, when the holes are so close together that D is much less than λ, the Huygens construction shows that the two holes act as if they had merged into one; then the outgoing wave is indistinguishable from a simple spherical wave without any interference fringes at all.

6

Quantum beanbags

6.1 A muddy wheel

After this excursion into the laboratory, let us return to the discussion of the equation of quantum motion. As in the case of classical mechanics, we start with the Descartes approach: what does a quantum do in the absence of forces? A classical particle moves with constant velocity when there are no forces; what does a quantum do? Picking out just one path for the quantum to follow would offend against the uncertainty relations. In fact, we must do exactly the opposite: each path that is not explicitly forbidden, and that is not distinguished from other paths in any way, must be allowed as a possible alternative.

The mere fact that a path traces a given curve through space and time is not enough to distinguish it from a differently traced alternative. After all, as long as the quantum does not interact with others along the road, there is no way to determine where it has been (this is the essence of the whole preceding chapter). Interactions along the path correspond to the appearance of forces, and these we had pointedly excluded; the action of a force is a way to distinguish among paths. All force-free paths are to be considered collectively as equal-rights alternatives.

The properties of waves allow us to assume that *any* line between two points is a possible wave path; we saw this when we discussed Huygens's principle and the tracing of wave paths using black-and-white phase tape. To carry this construction a little further, let us imagine that, in the electron interference experiment described in the preceding chapter, we add another screen with holes. Then the number of alternative routes from hole to hole increases (Fig. 6.1); and if we increase both the number of screens and the number of holes per screen indefinitely (so that the screens are entirely eaten

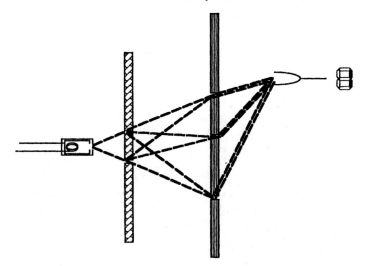

Fig. 6.1 Alternative paths which pass from source (left) to detector (right) through two screens. The paths have been marked with phase tape, indicating the phases of the various alternatives. At the detector, all amplitudes are added.

up by the holes), we obtain a limit in which any path from source to detector is possible, even if it went to Sirius and back.

The task before us now is the following. First, fix the beginning and the end of a quantum path (corresponding to the electron source E and a detecting counter C in the preceding chapter). Second, construct all paths from E to C. Third, mark each path with phase tape of constant wavelength. Fourth, determine the phase of each path at C, and find the corresponding amplitude. Fifth, find the total amplitude by linear superposition of all path amplitudes. Sixth, take the square of the total amplitude, to finally obtain the probability of finding the quantum at C.

These steps are a cookbook version of the wave superposition we have discussed before. Some remarks: one, we are allowed to use phase tape of constant wavelength, because in the absence of forces (i.e. in the absence of interactions with other quanta) the energy and the momentum are constant, so that the quantum frequency and wavelength are constant as well. If we did not require this, we would get immediate disagreement with conservation laws that we know to hold in the large-scale world. Two, when the phase tape crosses the detecting counter C, it has traced a whole number of phase cycles plus a little bit. The bit that remains is the phase at C. It does not matter how often the phase completes a whole cycle, because wave behaviour is periodic: after a whole phase cycle, the wave vector comes back to exactly where it was before. Three, it is extremely important that we should have *linear* superposition. If we did not demand this, we might be able to disentangle the superposition.

The linearity of an expression $a + b$ ensures exactly equal rights for its constituents a and b; thus, it is impossible to determine whether it was obtained by first taking a and then adding b, or the other way around. Consequently, we cannot obtain information about which path the quantum followed by taking a superposition apart, as long as the superposition is linear. If I give you the number 25, you could not say whether it was obtained from $17 + 8$ or from $2 + 23$. But if I tell you that it is the sum of two squares, you know that it must have been $9 + 16$. In this way, complementarity is safeguarded; if we used anything other than linear addition, there would be something else that distinguishes among the alternative ways of getting from E to C.

The steps indicated above can be performed algebraically, but that is a somewhat messy job. Instead, we can use a graphical process that illustrates the role of the phase quite well. As we saw in Chapter 2, uniform circular motion can be used to produce a wave. We now use this mathematical connection between waves and spinning circles to our advantage, imagining that we have a wheel with a circumference equal to the wavelength of our quantum. Let the source E be a small muddy puddle; run the wheel through E, and roll it anywhere you please (Fig. 6.2) until you cross the point C.

The wheel will leave mudprints all along its track, thereby tracing out the phase along the path of the quantum. At the detector, we note the position of the mud spot on the rim of the wheel. An arrow drawn from the axle to this spot represents the amplitude of the path that the wheel has followed. Such an arrow can be described algebraically by giving two numbers, say (p, q). A convenient way of doing this is by combining these two into a `complex number` $p + iq$. Accordingly, the arrow is sometimes called the `complex probability amplitude`. †

The linear addition process for vectors is simple: just link them all, tip to tail, in any order whatever. The fact that the order of addition doesn't matter is crucial: any other recipe would require a specification of what alternative comes first, but this would invalidate the requirement that the

---

† The number i is defined by $i^2 = -1$. Complex numbers are added just like vectors, namely by adding their components separately: $(a + ib) + (p + iq) = (a + p) + i(b + q)$. Complex numbers are multiplied by writing out the product: $(a + ib) \times (p + iq) = ap + iaq + ibp + ib \times iq$, which works out to $ap - bq + i(aq + bp)$. Thus, the square of a complex number is $(p + iq)^2 = p^2 - q^2 + 2ipq$. The square of the absolute value of a complex number is obtained by reversing the sign of i in one part of the product: $(p + iq) \times (p - iq) = p^2 + q^2$.

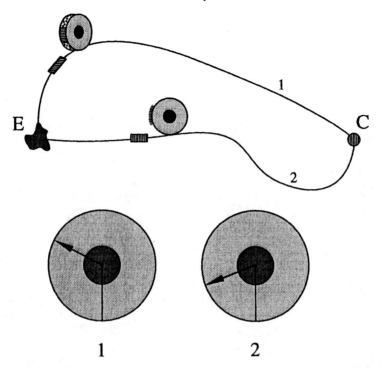

Fig. 6.2 Two different paths (alternatives) from the electron source E to the counter at C. The phase wheel leaves mudprints along the path. Each print indicates the completion of a full phase revolution, 360°. The amplitude vectors corresponding to these two alternatives are shown below.

alternatives are indistinguishable. Dropping this requirement is tantamount to dropping the uncertainty principle.

By adding them tip-to-tail we obtain a chain of arrows; the arrow drawn from the beginning of the chain to its end is the linear sum of all the arrows (Fig. 6.3). The square of the length of this arrow is the `total probability` of getting from E to C. For future reference, we should note one important property of this superposition recipe: the only things that count are phase *differences*. When you follow the above construction closely, you will notice that it does not matter if you give each muddy wheel at C an extra turn (say 271°), as long as you do this with all of them. The net result is merely that the final arrow is turned by the same amount (271° in the example), but its length will stay the same. Thus, the total probability does not depend on phases but on phase differences.

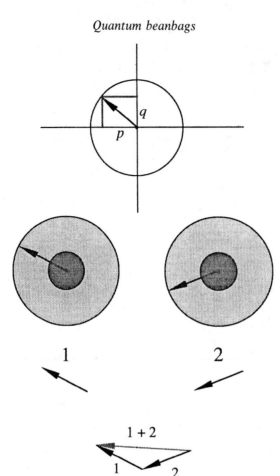

Fig. 6.3 Addition of the two amplitudes from Fig. 6.2. Amplitude arrows are added by laying them tip-to-tail, and connecting the beginning of the first arrow to the tip of the last. Instead of specifying the length of an arrow and its direction (technically: amplitude and phase), we may also specify the arrow's projections (components) p and q in two arbitrary directions, as shown at the top. As the arrow revolves, p traces a sine and q a cosine curve.

6.2 The importance of having phase

At last we are in a position to deduce which path corresponds to Descartes's inertial trajectory, the classical straight line of constant velocity. Of course, we cannot single out one particular path, because we saw that *all* tracks of the muddy wheel (including those that loop the loop around Sirius) contribute equally to the total probability. But we can ask: *which path is the*

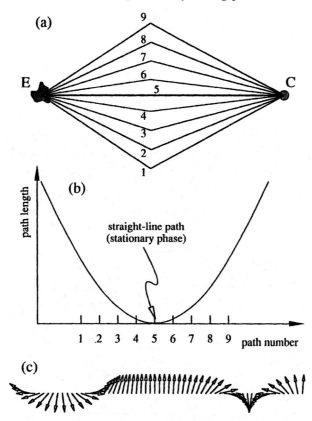

Fig. 6.4 (a) Triangular paths connecting electron source E and counter C. These alternatives are but a small special subset of all possible paths. (b) A graph of the path lengths and the corresponding amplitude arrows. Notice that, near the shortest one (number 5), the paths have almost the same length. Consequently, the amplitude arrows (c) corresponding to these paths point approximately in the same direction.

most likely one? Because of the existence of phase, there is a most probable one.

For the sake of simplicity, I will only consider a small subset of all possible paths in the diagrams, but you can verify that what follows holds for other paths, too. I have plotted the total length of paths that make wider and wider excursions into the Universe (Fig. 6.4). At first, the path length does not differ much from the straight-line distance EC, but gradually it increases out to Sirius and beyond. Next, I have plotted the phase (at C) of all these paths. We see immediately that, as long as the path stays close to the straight line connecting E and C, the phase changes hardly at all: we have a *stationary phase*.

When the paths differ more and more from the line EC, the phase begins to sweep up and down more and more. When we look at the way in which the total amplitude is built up from the contributions of all these paths (Fig. 6.5), we see that only the paths with stationary phase contribute effectively, so that the largest probability is associated with a narrow bundle of paths close to the straight line EC. We thereby find the connection between the force-free motion in classical mechanics and in quantum mechanics: the trajectory with constant velocity (the inertial trajectory) is the one with largest probability. In the absence of forces, the straight and narrow one connecting source and detector, centred on the classical constant-velocity trajectory, is the one that is most likely to be followed by the quantum.

Only along paths that are very close to the straight line between source and detector is the phase practically the same. All other paths are possible; but if they are not close to that straight line, they will have wildly different phases, so that their contributions cancel out. So it is possible for quanta to go all the way to Sirius and back, but their contributions cancel against those that go, say, to Arcturus and back.

For good measure, let me emphasize again that I have not said that the quantum is a wave; but it behaves like one, to the extent that each quantum alternative (i.e. path) has an amplitude vector with a certain phase. Phase is a bookkeeping number that tells us how many times a given length fits along a given path; the rotation of the amplitude arrow of each path is described by the phase. Upon superposition, the existence of phase ensures that some paths (those with stationary phase) contribute more to the total probability than others.

The path of stationary phase is the straight line that connects the source and the detector of the quantum. That is straight enough, but how narrow is it? In other words, how far towards Sirius do we have to stretch the alternative paths before quantum interference starts to kill their contribution? The length scale that determines the phase is, as always, the De Broglie wavelength λ, which is connected to the momentum p of a particle by $\lambda = 2\pi\hbar/p$. Accordingly, we find that the alternative paths practically cancel each other's contributions if they deviate more than a few times λ from the classical straight-line path. We saw before how small the De Broglie length can be: 10^{-28} metres for a falling grain of sugar, which is a million million million times less than the size of an atom. An object as large as a sugar grain stays that close to its Cartesian inertial path, so quantum effects can hardly be blamed if you spill the sugar over the rim of your cup.

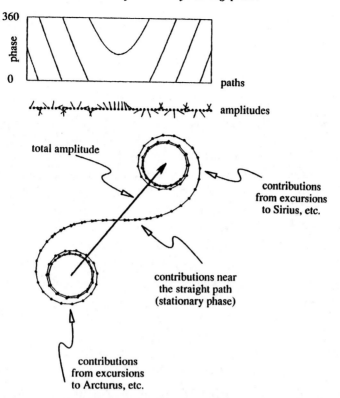

Fig. 6.5 Addition of amplitudes along many paths. The amplitude arrows, put tip-to-tail, form an elegant S-shape. The stationary phases near the shortest-line path have arrows which point almost in the same direction. Thus, these amplitudes add to a significant net result (the straightish section of the S). If we added infinitely many arrows, the net result would be one arrow, from one curl of the S to the other.

It seems very strange that paths which go via Arcturus or Sirius do, in fact, all contribute to making a particle go in a straight line. After all, doesn't a free particle go straight anyway? What is it that would stop it from doing so? But the point is that, due to quantum uncertainty, a particle is no longer free if we place a barrier *anywhere* in the Universe! Let us see what happens if we remove all possible excursions to faraway places. We see immediately that, if we block all paths except a few (for example by means of a screen with a hole in it), all points which lie on a circle centred on the hole have nearly the same phase. Thus, on this circle all the amplitude arrows add to roughly the same value, and *the most probable path spreads out into a spherical wave front* (see Fig. 3.6 or 3.12). Squeezing a quantum by removing alternative paths makes it spread out away from the classical straight line!

You can see for yourself what the consequences are of organizing phases. Take a lens from a camera, or a lens from spectacles with a positive correction strength. Make an image of some object, preferably a sharp light source such as a distant lamp. Now block the lens in random places by sticking little bits of masking tape on it. You will see that the image projected by a random-masked lens is the same as that of an unmasked lens, only fainter: the lens surface is effectively smaller because of the tape. But now block the lens with a *regular* structure, such as mosquito netting or wire mesh cloth: you will get spikes and rays in the image, even though the total blocked area is minimal. This is because you have *systematically* cut out certain paths. Thus, the corresponding amplitudes have been removed systematically, too: only certain specific phases are now lacking. That means, that in the overall amplitude sum we obtain a systematic effect. Regularity plus phase gives orderly interference.

6.3 Feynman paths

To summarize the discussion so far: when we consider the world on a large scale we find that, in the absence of forces, a particle moves with constant velocity. When we look at the world of the very small we find that, in the absence of forces, the most likely path of a quantum between two points is the path of stationary phase, which corresponds to a straight-line path with constant speed. This does not seem particularly exciting: we obtain straight-line motion with constant speed in both cases. At least we have shown that the quantum rules reproduce what we see on a gross scale. But how are we going to include forces in this picture? What is the quantum equivalent of the classical equation of motion?

To solve this question, we note two very important things. First, the essence of finding the most likely quantum path is the calculation of the phase, a number that tells how many times the De Broglie length λ fits along any path. Second, by the quantum rules, λ is inversely proportional to the momentum (or, alternatively, the velocity) of a particle. In classical mechanics we saw that the effect of a force is a change of momentum (you may recall that the total momentum of a system on which no external force acts is constant in time). Accordingly, it seems sensible to assume that *the effect of a force is a change of phase*. Thus, not all paths are equal any more: we add an extra something to a trajectory through space and time, and we

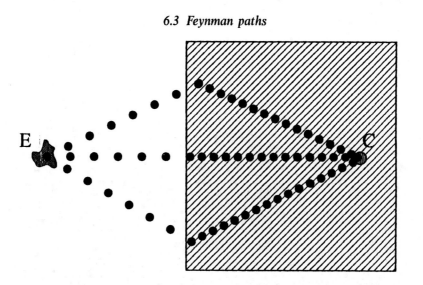

Fig. 6.6 Paths for the Snellius experiment. Shown are the mudprints of a phase wheel that changes diameter. In the shaded zone, the wheel is smaller so that the prints are closer together.

express the influence of that something by specifying its effect on the phase of the path.

Can we deflect a particle from a straight-line path by tinkering with its phase? Yes – but the general case is mathematically messy, so I will describe an example that will turn out to be very familiar (Fig. 6.6). Consider again a particle source E and a detecting counter C. Suppose that half-way between E and C we place a surface where the momentum of the particle changes. For definiteness, I will assume that this reduces the De Broglie length of the particle from 1 to 0.5. Now we ask: what is the path of stationary phase?

We roll our muddy wheel from E to an arbitrary point X on the momentum-changing surface. At X, we change wheels to one with half the diameter of the preceding one, and roll on towards C. Without the change in momentum, we find that the phase reaches a minimum (i.e. a point of stationary phase) at exactly that position X where the straight line from E to C crosses the surface. Of course: we have already seen that, in the absence of forces, this straight line is the most probable path.

But with the change of momentum taken into account, we find that the phase becomes stationary when the point X is *not* on the straight line EC! Thus, we see (Fig. 6.7) that a change in the De Broglie wavelength at X produces a change of phase which causes a kink in the most probable path of the particle. In our large-scale world, when something gets knocked sideways

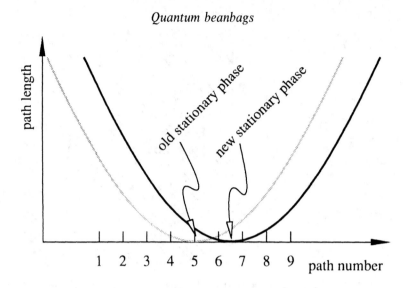

Fig. 6.7 Path lengths for the experiment in Fig. 6.6. Notice that the point of stationary phase has *shifted*. Formerly, the most probable path was the straight line between E and C (path number 5). Here, the stationary-phase path is somewhere between 6 and 7. This path is no longer straight, due to the phase shifts in the shaded area of Fig. 6.6.

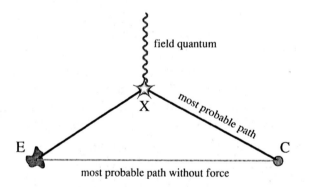

Fig. 6.8 Change of phase at an interface interpreted as interaction with a field quantum.

we say that a force acts on it. In the quantum world, the force is mediated by a particle (Fig. 6.8) that transmits the force by direct interaction.

Incidentally, the above example illustrates what we call refraction in classical physics: the kink at X, and the bent path EXC, corresponds to the apparent break in a stick poking through a water surface. The kink at X is described by Snell's law, which in fact can be rigorously derived from this type of analysis.

This simple example shows that it is indeed possible to deflect a particle from its inertial path by varying its phase: the path of largest probability is bent, which corresponds to the classical-mechanics statement that a force has acted on a particle. But what prescription should we adopt for the way in which the phase depends on external influences?

Because of the De Broglie connections between momentum and wavelength, and between energy and frequency, we suspect that energy and momentum will play a central role. Moreover, whatever quantum path description we devise, we must at least require that (on a large scale) it reproduces the results of classical mechanics. This reinforces the suspicion that phase and energy-momentum must be intimately connected. Nature has provided us with these clues, but that is not enough to deduce the phase prescription strictly. Indeed, we cannot derive laws of Nature as we do with mathematical theorems, but, rather, we must infer them through some creative insight. The crucial discovery about quantum phases was made in 1942 by Feynman, who linked the phase of a given Feynman path with a particular form of energy, called the action, associated with that path.

In classical mechanics, the action (usually called S), is equal to the kinetic energy *minus* the potential energy, summed over the time-of-flight of the particle. One can derive all the usual laws of classical motion by stating that a particle follows the path where the action is stationary. That is to say, if we were to calculate the action S associated with the true path, and also the action S' of a path adjacent to the true path, we would find that S' is very nearly equal to S, in such a way that S is an extremum (i.e. maximum or minimum) on the true path. This property of *stationary action* suggested to Feynman the equality (phase) = (action)/$\hbar$. In the Feynman picture, the quantum of action, $\hbar$, is the amount of action needed† to shift the phase by one radian, which is $180°/\pi$, just a little over $57°$.

Thus, stationary action (which leads to the correct motion in classical mechanics) corresponds to stationary phase (which produces the quantum path of largest probability). When discussing linear superposition, we saw that the final probability associated with quantum motion depends only on a phase *difference*; it is then natural to write the phase prescription in the form of a phase *increment*, namely

† I strongly recommend that you read Feynman's own account in *The Feynman Lectures on Physics*, Vol. II, p. 19-1. Notice, by the way, that Planck's constant $\hbar$ is included in the expression of the phase only as a sop to the older literature. From a physics point of view, carrying $\hbar$ along is a pure waste of effort, because we can choose different physical units to make $\hbar$ equal to 1, by definition.

(change in phase) $=(1/\hbar)\times$(change of action along the path)

With every small step along the path, the action changes a little; thus, the action is proportional to the step size. The proportionality factor is called a `field density`, and we obtain

(change in phase) $= (1/\hbar)\times$(field density)$\times$(step along path in space-time)

In the Feynman path picture, the classical equations of motion are replaced by the following prescription. If you want to find the probability that a quantum starting at a point E reaches a detector at a point C, you must first identify all space-time paths that are not explicitly forbidden (we will consider forbidding rules presently). Second, calculate the field density wherever these particles go. Third, find the phase for each path by summing up the phase changes as you go along from E to C. Fourth, add the amplitude vectors (arrows) using the proper phase, in linear superposition. Fifth, the square of the length of the total amplitude vector is the desired probability.

6.4 Superposition and quantum sorcery

The entire mechanism of Nature has thus been written down as the superposition of Feynman paths. It does not seem to be so difficult to pick paths in space-time, but what is the field density going to be? At this point, we make the connection with our earlier discussion of Feynman diagrams and the exchange of quantum beanbags. As I have stressed repeatedly, a quantum field is not continuously variable like a classical field, but instead must be thought of as a superposition of field quanta (the replacement of a continuous field by a population of quanta is sometimes called `second quantization`; Fig. 6.9). Field quanta can be included in the Feynman path picture if we choose the points E and C to be vertices of an exchanged field quantum. Accordingly, the paths are replaced by an infinite web of vertices connected by exchanged quanta; the behaviour of Nature should then be calculable if we can specify what the possible quanta are, and how they are coupled together.

So let us return to the description of the quantum beanbags that mediate forces by being exchanged between vertices. You may recall that in a preceding chapter we asked ourselves: what's in the bag? The beans in the bag represent information of some sort which is exchanged between quanta that are subject to the force mediated by the bag.

The existence of an absolute scale of smallness, and the consequent quantum behaviour as expressed by the uncertainty relations, implies that there is a limitation on the amount and precision of the data carried by the bag. We

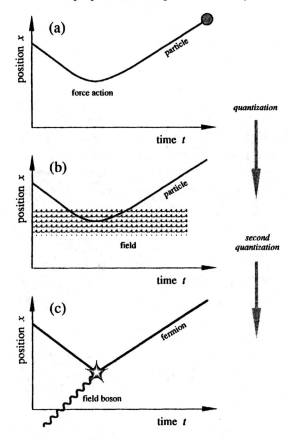

Fig. 6.9 Second quantization. (a) The classical view, in which the path of a particle is bent by a force; notice that the force is mysterious, and is invoked from nowhere. (b) The 'first' quantization, in which a quantum particle interacts with a field. Here, too, the field is dragged in from nowhere. (c) The picture after 'second' quantization, in which the field, in its turn, is interpreted as consisting of particles which interact directly with the incoming particle and knock it sideways. All particles are treated on the same footing – except for the boson–fermion distinction – and the field is not added arbitrarily.

will call a quantum beanbag with a specific set of data on board a `state`. Each exchanged quantum then carries a set of labels, one for each data bean. Such a string of labels associated with a quantum is often called a `state vector`. Because of the absolute scale of smallness, we expect that states become increasingly simple as they become smaller: states should contain fewer and/or more indefinite data as we descend the subatomic distance ladder. (A word of caution: as we will see towards the end of our story, it is found experimentally that more and more new varieties of databeans keep popping up as we look at smaller and smaller beanbags. This baffling behaviour has so far prevented us from reaching a scale of smallness where simplicity reigns).

Suppose, then, that we are given a specific quantum state in the form of one particular state vector. We saw above that new states can be built up by superposition of others. Hence it is perfectly natural to ask: given a state, what is it a superposition of? In other words, what is the decomposition of a state? The generality of the superposition principle leaves us no choice but to admit that a state can be decomposed into a possibly infinite number of other states (Fig. 6.10). Any state is a quantum chimaera, which may be considered simultaneously as one thing and as another (in Greek mythology, the Chimaera was a composite beast with the head of a lion, the body of a goat and the tail of a serpent). In the quantum world, we can expect to give a cogent answer to the question how many percent of a raven is like a writing-desk. Any decomposition of a state is allowed (but only a few may be useful for practical calculations).

When a state is formed by the superposition of two others, the resulting chimaera has properties that are intermediate between those of its constituent states. By 'intermediate' I refer expressly to the outcome of possible interactions with that state; in this way, a link is made between superposition and indeterminacy, as follows. Given a quantum state called L. Suppose that a certain measurement, performed on L, yields with absolute certainty the quantity l as outcome (note that the uncetainty relations never forbid the precise knowledge of a *single* quantity). Given a state G; suppose that the same measurement, performed on G, yields with certainty the quantity g. We say that L and G are eigenstates ('pure' states) of that particular measurement, with eigenvalues l and g. If a state C is a superposition of G and L, then C is called a mixed state. What is an eigenstate depends on the type of observation that is being performed; thus, a given state may be an eigenstate in one respect but not in another.

Now make a state that is a superposition of L and G. What will the measurement yield if performed on it? The result will sometimes be l and sometimes g, but *never* anything else. The intermediate nature of the superposed state shows up in the *probability* with which result l or result g emerges from the observation, and not in the result *itself* being intermediate between l and g. In this way, superposition and probability (hence indeterminacy) are inextricably intertwined.

As an analogy, consider a quantum sorcerer's apprentice who has been given the task of constructing a chimaera. Suppose that this is undergraduate

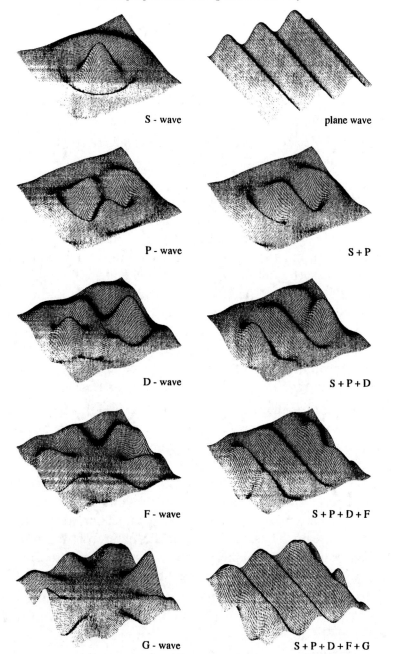

S - wave

plane wave

P - wave

S + P

D - wave

S + P + D

F - wave

S + P + D + F

G - wave

S + P + D + F + G

Fig. 6.10 Building up a plane wave by superposition of spherical harmonics. The picture on the top right-hand side shows a plane wave. On the left, we see a series of five wave states such as occur in atoms. These are called 'spherical harmonics' and are indicated by various letters: S for the wave with one central maximum, P for one with two maxima, then D, F, G, and so forth. Below the plane wave, we see the result of adding successive spherical harmonics together. The more such waves we add, the more the result looks like the original plane wave. Thus, a plane wave consists of a certain percentage of S-wave, another percentage of P-wave, yet another of D-wave, and so on. Tricks of this type are used commercially, for example in CAT scanners.

sorcery, so that only a two-component chimaera is required; a lion–goat superposition, say. The apprentice captures a pure lion state L, puts it into a cauldron with a pure goat state G, and starts stirring. After having swished the mixing ladle a certain number of times around the pot, the chimaera is declared done and is taken out for inspection. To what extent is it a lion, and to what extent a goat?

This question is decided by experiment. The chimaera is offered a cabbage and a zebra steak. If it bites the cabbage, it is goatlike; if it bites the steak, it is lionlike.

In the professional literature, one would read that the chimaera is 'in a goat eigenstate' in the first case, and 'in a lion eigenstate' in the second case. Obfuscators would say that the chimaera collapses into a goat or a lion state, as the case may be. But there really is no such thing as the 'collapse of the wave packet': when you detect a particle in any state, all you know is that it is in that state; as Popeye said, 'I yam what I yam'. You know *nothing* of 'what it was before'. Indeed, your measurement is perfectly consistent with its having been in that state ever since the Big Bang. But the better you know that state (= the past), the less you know the future. Compare the electron experiment in the preceding chapter: the narrower the aperture, the more the outgoing wave looks spherical, i.e. structureless.

If the chimaera bites the cabbage, it is goatlike, and if it bites the steak, it is lionlike. The experiment is performed a large number of times, and lo! the animal sometimes bites the cabbage, sometimes the steak, but *never* does it attempt to bite both at once. Moreover, it proves to be impossible to predict, on any given occasion, whether the chimaera is in the goat eigenstate G or in the lion state L. But a long series of tests allows the apprentice to say with great confidence how probable it is that the chimaera will choose the cabbage or the steak. Thus, the intermediate nature of the superposition $L + G$ shows up in the probability that the result l (the chimaera bites the steak) or the result g (it bites the cabbage) emerges from the test.

If you detect a particle in any eigenstate, all you know is that it is in that eigenstate. The more certainly you know that the particle is in that state, the more precisely do you know its past. However, this implies that you know correspondingly less about the particle's future. This can already be seen in our preceding discussions of waves passing through apertures. When the opening is very small, the outgoing wave is spherical; which is to say that yes, you know precisely where the particle went, but no, you haven't a clue about where it will go. Thus, there is no need to talk about 'collapse of a wave packet'. The observed state (an electron, Schrödinger's cat, or whatever) is what it is; but before you have interacted with it, you do not know it yet.

In conclusion, let us consider how the superposition $L + G$ depends on the number of times θ the ladle is stirred around. Of course we must demand linear superposition, as before; thus, if g is the goat amplitude and l the lion amplitude, we require the chimaera amplitude to be $lS + gC$, in which S and C are numbers that depend on θ in a way to be determined presently.

Our first requirement is that the chimaera either bites the cabbage or the steak, and never anything in between. The probability P is the square of the amplitude, so we find that $P = (lS + gC)^2 = l^2 S^2 + g^2 C^2 + 2lgSC$. The first term is the probability that the chimaera is lionlike, the second term that it is goatlike, and the third term that it attempts to bite both the steak and the cabbage at once. Thus, we require that the average of SC is zero: in that case, we have the correct quantum behaviour.

Second, we require that the superposition of a goat and a goat should always be a goat: $gS + gC$ should correspond to a pure cabbage-eater. Because we have already required that $SC = 0$ when averaged over the stirring angle θ, we obtain $P = (gS + gC)^2 = g^2(S^2 + C^2)$. This must equal g^2, and therefore $S^2 + C^2 = 1$. There are two mathematical functions of θ that fulfil these requirements, namely $S = \sin\theta$ and $C = \cos\theta$. It is easily verified that $\sin^2\theta + \cos^2\theta = 1$ and that $\sin\theta \times \cos\theta$, when averaged over θ, equals zero. By the way, there are more complicated mathematical objects which can be used like sine and cosine in a mixing recipe such as this. You may think about these as some kind of superrotation. But the example introduced here is complicated enough (just pronouncing 'chimaera' is a bit of a twister already).

Thus, the chimaera amplitude is $l\sin\theta + g\cos\theta$. The angle θ is, most appropriately, called the `mixing angle`. For example, if the mixing angle is $\theta = 25°$, we have $\sin\theta = 0.423$ and $\cos\theta = 0.906$. This particular chimaera would bite the cabbage in 82.1% of all trials, and would bite the steak with a probability of 17.9%.

Are the functions $\sin\theta$ and $\cos\theta$ the only ones that produce proper superpositions? No – in fact, there is a whole family of functions which give $s^2 + c^2 = 1$ and an average $sc = 0$, but they are all closely related to the sine and cosine functions, so we will continue to use these as examples.

Using superposition, we can understand quantum motion from a somewhat different point of view. Let us describe a quantum at a given spot in space by the state $|H>$ (for Here), and let the same quantum at another location be described by the state $|T>$ (for There). Then a quantum that moves between these two spots is naturally described by the superposition $x|H> + y|T>$.

The quantities x and y are numbers that indicate the change of probability over the course of time. The changeover from $x = 1$ to $x = 0$, and an associated change of y from 0 to 1, then describes the motion from here to there. It turns out that x and y depend on the *action* we have encountered above.

6.5 The embezzlement model

If you have read the beanbag-throwing analogy carefully, it will not have escaped your attention that the mechanism as it stands is clearly impossible. Let us consider a single vertex again, and scan it with the Feynman diagram scanner. At the vertex, our hypothetical skater throws the bag; but that supposes that the thrower has an internal store of energy. Even if a quantum had such a store, it would ultimately run down, and we cannot very well propose an electron that eats in order to maintain its throwing arm. Without an internal energy supply, energy and momentum are not both conserved at a single vertex. We can arrange the recoil so that momentum is conserved; but then energy cannot be, since the recoiling particle *and* the thrown beanbag have a non-zero velocity with respect to the location in space where the throw occurred.

At this point, the quantum behaviour of the Universe comes to the rescue. We must pose our problem a bit more subtly. The question is not: 'is' energy conserved at the vertex, but rather: can we *observe* that energy is not conserved? In order to observe a violation of energy conservation, we would have to intercept the virtual quantum. Because a quantum is either wholly absorbed or wholly not absorbed, intercepting it would destroy the interaction. So *either* we let the interaction do its thing, in which case we will never know firsthand about the doings of the virtual particle; *or* we capture the intermediary beanbag, accepting it as a real particle, and killing the interaction. Notice the similarity with the experiment which I set up to determine through which hole an electron passes!

It turns out that the above can be stated in a more quantitative way. In our discussion of the De Broglie rules and the Heisenberg uncertainty relations, we concluded that the uncertainty in the energy of a quantum, multiplied by the uncertainty of the time at which the quantum is caught, must exceed Planck's constant. This uncertainty relation implies the following: a virtual quantum can violate conservation of energy by an amount that is no bigger than Planck's constant divided by the lifetime of the quantum. If the energy debt is paid back quickly enough, we will never be able to construct an experiment that catches the quantum's transgression; if we were to try and measure the energy imbalance, the very fact that we perform the experiment upsets the energy balance more than the virtual quantum itself would.

Metaphorically, a virtual particle is like an embezzler who takes a sum of money out of a bank. Suppose the bank's computer sets off an alarm as soon as the interest on unaccounted-for capital has exceeded a fixed sum, say 50 guilders (it wouldn't be able to track every cent exactly, due to roundoff errors). If the interest rate is 10% per year, an embezzlement of 1000 guilders would have to be paid back within half a year to go undetected, whereas a million has to be back in four hours if the crime is to escape notice. Imagine that the bank mounts a campaign against embezzlers. Because the big criminals operate so quickly, they are very elusive, so the cost of such a campaign would increase proportionally to the magnitude of the embezzlement. Any such investigation upsets the accounting of the bank more than the embezzlement it was supposed to counteract.

Thus, we are led to the Embezzlement Model of quantum interaction: 'anything goes, as long as it escapes detection'. If this were really true, we would expect the world to be utter chaos, whereas the merest glance at a crystal or a flower proves that it is not all that bad. There must be at least some things that are more probable than others. But which? And how do we formulate any special preferences that Nature might have for what happens at the vertex? Once we know how vertices behave, we know how the forces behave that govern our Universe.

7

Symmetries

7.1 A forbidding prospect

The essential uncertainty of quantum behaviour might create the uneasy suspicion that anything goes in this world. But since the Universe has (on all observed scales of length, time, mass and so forth) a very definite structure, it seems extremely unlikely that everything is allowed. There must be at least some things that are more allowed than others.

As it happens, there are many things that are forbidden (in the sense that they are unlikely in the extreme), and it is the forbidding rules that give structure to the world. We will discuss quite a few of these rules, and it is important to bear in mind that the origin of these is essentially not understood. It is not known why the rules must be as they are; some day we may attain a deeper level of understanding, which will provide insight into the rules that govern the rules, as it were. But we are not at that stage yet, and for the moment all rules that have been discovered seem arbitrary and highly non-obvious. If you were to design your own universe, it might not occur to you to do it the way our world is, and we know of no reason why you should impose all the experimentally observed rules in order to obtain a working universe.

If you wanted to organize a universe, you could compile a whole book full of rules, a long list of entries that say 'Thou shalt not this', 'Thou shalt not that'. But if your universe consisted of a possibly infinite multitude of inhabitants, which could do an infinite multitude of things at an infinite number of points in time and space, then such a rule book would surely be multiply infinite. If you were to write, in this way, a complete scenario for your world, the universe you would be trying to organize would be too small to contain all the rules that are supposed to govern it.

blue   red

(a)

(b)

Fig. 7.1 The appearance of spectral series is not random at all, which led to the use of light spectroscopy as a means for probing the inside of atoms. (a) Sequence of lines placed randomly. (b) Hydrogen spectral lines.

Alternatively, you could lump together a large (possibly infinite) number of rules in a single all-encompassing one. Thus, you might write in the Constitution of your universe, 'Do unto others as thou wouldst others do unto thee', or even more briefly, 'Love thy neighbour as thyself'. Such an overall rule, that summarizes a multitude of individual possibilities, is called a symmetry. The symmetry just mentioned is between you and your neighbour. You are surely familiar with social symmetries: you and your neighbour have the same rights, even though you are certainly not the same, and may differ as to age, race, sex, creed, ethnic origin, and what not. It turns out that the symmetries that govern the quantum world are of a very special type, called symmetry group by mathematicians. I will mostly restrict the discussion to this kind of symmetry.

As an example of the way in which the number of rules required to specify something can be drastically reduced, consider the spectrum of hydrogen (Fig. 7.1) discussed in Chapter 2. When presented with a plot of the positions of the main spectral lines of the hydrogen atom, we suspect immediately that the lines are not distributed at random. But in the absence of any overall formula, we can only describe the spectrum by listing its lines one by one. Then someone notices that there is a certain regularity in the distribution of the lines: it is possible to group them together in orderly progressions (Fig. 7.2). However, the relationship between these series is still obscure, until one happy day it is discovered that the whole spectrum can be reproduced by the following recipe. If E is the energy† of a spectral line, then $E = R(1/k^2 - 1/n^2)$, where k is the serial number of a series, n is any integer larger than k, and R is a constant number.

† You may remember that energy and frequency are related by $E = 2\pi\hbar f$; since frequency and wavelength are related by $c = f\lambda$, we obtain $E = 2\pi\hbar c/\lambda$.

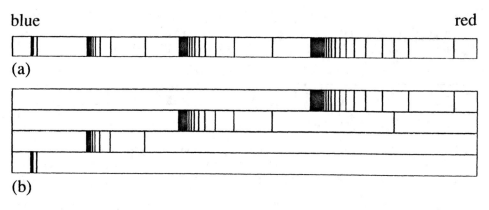

blue   red

(a)

(b)

Fig. 7.2 The hydrogen spectrum of Fig. 7.1, with the series of lines separated out. From bottom to top: Lyman, Balmer, Paschen and Brackett series. Topmost spectrum shows the sum of all these series.

Presto! What economy of means in such a simple equation. Not only do we find an underlying regularity, so that we can describe the known spectral lines by means of two `quantum numbers` called k and n, but we can use this expression (called the `Rydberg equation`) to predict where other spectral lines may be found. And indeed, lines are observed at the predicted positions.

Even though the Rydberg equation captures the hydrogen spectrum in a simple expression, the regularity it produces is not precisely the kind of symmetry group with which we will be mainly concerned. However, the Rydberg formula is closely related to such a group: we will show in Chapter 9 that it can be derived from the requirement that the equation which describes the hydrogen atom must be symmetric under rotations in space. It will then turn out that the rotational symmetry group imposes the shapes that a hydrogen atom can assume, and that the energies associated with these structures are accurately reflected in the hydrogen spectrum.

The above sequence of events closely resembles the history of high-energy particle physics. At first, we are presented with a bewildering array of particles: they have different masses, electrical charges and interaction behaviours. Then someone makes an inspired guess about which particles belong together, in some sense: they are arranged together in `multiplets`. How elementary or fundamental these arrangements are remains to be seen. The particles in a multiplet, like the spectral lines in a series, are supposed to be related through the action of a symmetry group. Furthermore, it turns out that such groups have the wonderful property that they give precise prescriptions for the ways in which the particles in a multiplet can interact.

Symmetry forbids. Forbidding imposes order, but many different things that possess a certain order may derive from the same symmetry (e.g. all planets possess approximate rotational symmetry, but they have very different masses and compositions). That is why physicists believe that the underlying symmetry, which forbids whole classes of occurrences at one stroke, is, in a sense, more fundamental than the individual occurrences themselves, and is worth discovering.

7.2 Groups and rotations

Suppose that you walk down the street, and, arriving at a corner, make a left turn. You will observe that neither your displacement along the street, nor your turning, changes you. It is by no means self-evident why this is so. If you were to construct your own universe, maybe you could arrange things in such a way that a person who is displaced over a certain distance in a straight line changes into a white mouse, or that someone making a right-angle turn changes into a pumpkin. If that were the case, the expression 'turning into' something would really have forceful meaning! In subsequent chapters, we will see that rotation in another kind of space than our familiar three-dimensional space *does* turn particles of one kind into another. In fact, we already encountered an example of this behaviour when we discussed the sorcerer's apprentice concocting a chimaera: at a turn of the mixing angle, this beast could change its appearance from lion to goat and back again.

In our Universe we observe that things do not change when moved: mechanical systems are invariant under translation (moving down the street) and under rotation (turning a corner). Translation and rotation are symmetries of a very special kind: unlike loosely defined symmetries of the type 'I am as good at quantum mechanics as you are', or regularities like those expressed in the Rydberg equation for the hydrogen lines, translation and rotation are symmetry groups. A group is a set of objects, called group members or elements, plus some prescription for relating these elements to each other. I will show some examples presently, but it is important to define these prescriptions first, because it is precisely these interrelationships which make a group such a useful structure.

Group elements are related as follows: (1) With every pair of elements is associated a third one. The association is known as group `multiplication`, usually called multiplication for short, and is indicated by $\otimes$. Thus, from two elements a and b we can form $a \otimes b = c$. Note that we do *not* require that a, b and c are necessarily different, nor do we require that $a \otimes b = b \otimes a$. (2) The multiplication is `associative`; that is to say, if we chain three elements together by multiplication, it does not matter which of the two multiplications we perform first. In symbols, this means that $(a \otimes b) \otimes c = a \otimes (b \otimes c)$. (3) There is a unique element of the group, called the `unit element` (usually indicated by e or I), such that for each element a, one has $a \otimes e = e \otimes a = a$. (4) Each element a has an `inverse` (usually called a^{-1}) such that $a \otimes a^{-1} = e$.

You can easily verify that the integer numbers $(1, 2, 3, \cdots$, their negatives and zero) form a group under ordinary arithmetic addition. Just a little more complicated are the whole hours on a clock. They are also a group under addition, but here the number of elements is finite. The unit element is 12, because $h + 12 = h$ for any hour h (if we forget about a.m. and p.m.; otherwise, the unit element is 24). The inverses are the hours that are on equal and opposite sides of 12; on a clock, 11 is the inverse of 1 because $11 + 1 = 12$ (the unit element). Likewise, 10 is the inverse of 2, and so forth. Note that $6 + 6 = 12$, so that 6 is its own inverse.

In physics, there are groups of which the elements can be associated with some manipulation. Thus, each element of the `translation group` in space can be read as an instruction: 'Carry from here to there'. Likewise, each element of the `rotation group` corresponds to the instruction 'Rotate so much about this axis'. An object which remains the same when manipulated by a group element in this way is called `symmetric` under the group in question. A sphere is symmetric under the group of all space rotations; a cube is symmetric under rotations over 90°; a shoe is not symmetric (but a pair of shoes is, which shows a remarkable property of the space we live in). Often we find that an *approximate* symmetry is still useful; Earth is not quite a sphere, but it is close enough to allow accurate navigation.

Some groups can be shown to be equivalent to each other, in the sense that to each element (say p, q or r) of the one corresponds precisely one element (P, Q and R, respectively) of the other, and the products of the elements correspond likewise ($p \otimes q = r$ and $P \otimes Q = R$). For example, it is easy to see that the group of rotations over 30° in a plane is the same as the clock-number group mentioned above. Mathematicians have made complete classifications of groups, giving abstract definitions for each class. Often, such a class is named after its most famous member.

A specific instance of a group is called a `representation`; for example,

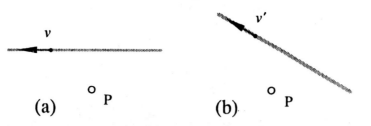

Fig. 7.3 (a) A particle moving with velocity v straight past an observer at P. (b) The same, but rotated over an arbitrary angle. The velocity has changed from v to v'.

the clock numbers are a representation† of the rotation group. In all cases that we will discuss, the representations have a finite number of elements.

The elements (that is to say, the members) of a representation form a `multiplet`, indicated in the professional slang by a name with a Latin root. Thus, a multiplet containing one member is called a `singlet`, with two members it is a `doublet`, and so forth. A sphere is a singlet under space rotations, a regular tetrahedron (which has four corners) is a quartet and a cube is an octet.

In the case of rotations in a plane, it does not matter in which order the rotations are executed. For example, you may turn 18° first and then 44° or vice versa, you will always end up with the same result. A group of that type is called `commutative` or `Abelian`. There is a theorem which says that all one-dimensional groups are Abelian.

7.3 Invariants and angular momentum

Every symmetry leaves something unchanged. In the case of rotations, that something is the distance between any two points, as prescribed by the `Pythagoras recipe`: $(distance)^2 = x^2 + y^2$, if x and y are rectangular coordinates. Such a quantity is called an `invariant` by mathematicians. Physicists sometimes call it a `conserved quantity`. For example, the fact that a particle does not change when it is displaced over some distance in a straight line is called `translational invariance`; it can be proven that, as a consequence, there is a mechanical quantity called `momentum` (i.e. mass times velocity) which is conserved. This property is entirely general:

---

† Finding representations can be a very tricky exercise. Usually, the representations are `matrices`; a matrix is a rectangular array of numbers that can be added and multiplied according to certain abstruse but useful rules. See, for example, *Lie Algebras in Particle Physics* by H. Georgi.

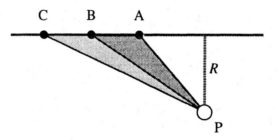

Fig. 7.4 The law of equal areas, as deduced from angular momentum conservation during straight flyby. In a given time interval, the particle moves from A to B. In the next time step, it moves an equal amount from B to C. The triangles PAB and PBC have the same base (because the speed is constant) and they have the same height R; therefore, their areas are equal.

a famous and extremely important result, Noether's theorem, says that *for each continuous symmetry there is a corresponding conservation law.*

Similarly, a particle does not change if it turns a corner: space is symmetric under the rotation group. According to Noether's theorem, there must then be a corresponding conserved quantity. In analogy with momentum, which is conserved because space is symmetric under translation, the conserved quantity that is associated with rotation is called angular momentum.

Consider an object with mass m, moving at constant speed v in a straight line (Fig. 7.3). Suppose you are standing at a point P; then you would see the particle fly by as shown at (a) in Fig. 7.3. Suppose, now, that we rotate the whole picture about P over an arbitrary angle. Then we obtain a view as shown at (b). What we are looking for is a quantity associated with the particle's motion that is the same in both cases. Clearly, the orientation of the system has changed, including the direction of the velocity and the coordinates of the moving object. But we remember that distances between points are conserved under rotations. Therefore, the speed v of the particle is the same before and after the rotation, because (speed) = (distance covered)/(unit time interval), and distances are invariant.

What other useful distances are there? Suppose we follow the particle along its path, and mark its position every second. Then we obtain the situation shown in Fig. 7.4. The distances PA, PB, and so forth, increase as the particle flies by, so clearly we cannot very well use these as invariants. But the triangles PAB, PBC, PCD, and any that follow, do have something in common: they all have the same height R, where R is the perpendicular distance from P to the particle's path. Consequently, the areas of these triangles are the same, and we conclude that the area swept out in one unit

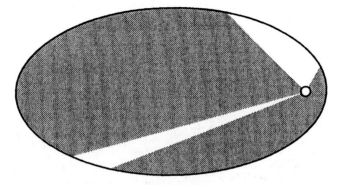

Fig. 7.5 Kepler's second law. If we follow a planet in its orbit, and if we pretend that the Sun (white circle) and the planet are connected by a line, then that line sweeps out equal areas in equal times, as indicated by the white zones. This is due to angular momentum conservation.

of time by the line connecting P and the moving particle is constant. You may recognize this as Kepler's second law (Fig. 7.5). Since that area is equal to $\frac{1}{2}R \times$ (distance AB), and because AB is equal to $v \times$(unit of time), we find that the constant area is proportional to vR. Thus, we expect that the mechanical quantity J that is invariant under rotations is proportional to vR, so that we can write $J = mvR$. The constant m turns out to be the mass of the particle, and J is the angular momentum.

It can be shown that J is still conserved under rotations if the orbit of the particle is not straight, provided that for R we take the perpendicular distance to the line that passes through the particle and lies in the direction of its velocity; we pretend, as it were, that the mass moves on a straight line for a very brief moment.

In particular, if a mass m moves with constant speed v on a circle with radius R, its angular momentum is $J = mvR$. Here we have a quantitative explanation of the skater's pirouette: if you start spinning with your arms stretched out sideways, R is large. Pulling your arms towards your body diminishes the distance R, but since the product vR is constant, v must increase to compensate: your rotational speed increases. And all this, mind you, because of a fundamental symmetry of Nature!

When we are dealing with an extended body instead of a point mass, we must take the body's orientation in space into account when describing its motion. In that case it is often very useful to consider the rotation of an object on its axis separately from its overall motion with respect to some external point. In the first case, we speak of spin angular momentum or

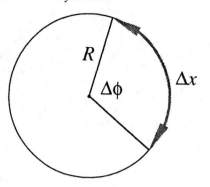

Fig. 7.6 Locating a particle on the circumference of a circle. The length Δx of the arc subtended by the angle $\Delta\phi$ is $\Delta x = R \times \Delta\phi$. Heisenberg's uncertainty relation can be cast in a form that connects the uncertainty $\Delta\phi$ with the uncertainty ΔJ in angular momentum: $\Delta J\,\Delta\phi > \hbar$.

spin; in the second case the name is *orbital* angular momentum. But this is purely a distinction of convenience; both arise from the same rotational symmetry of space.

7.4 Angular momentum and uncertainty

We have noted that the momentum and the position of a particle are complementary: they cannot be exactly determined at the same time. As we will see presently, it follows from this that *the angular momentum and the angular position of a particle are complementary, too.* Consider a particle moving at constant speed v on a circle with radius R (Fig. 7.6). If, from the centre of the circle, the particle is certainly seen within the angle $\Delta\phi$, then the particle is certainly on an arc segment of length $\Delta x = R\,\Delta\phi$. This expression gives a relationship between the uncertainties of the position and the angle.

Now the Heisenberg uncertainty product is $\Delta p\,\Delta x = m\,\Delta v\,\Delta x$, so that the relationship between Δx and $\Delta\phi$ leads us to conclude that $mR\,\Delta v\,\Delta\phi > \hbar$. Then we recall that $mvR = J$, the angular momentum. Therefore, we finally find $\Delta J\,\Delta\phi > \hbar$: *angular momentum and angular position are complementary*! This result must be true in each of the three dimensions of space, or else space could not be invariant under rotations.

This has some interesting consequences. First, let us suppose that we try to measure the angular momentum of a particle, for example that associated with an electron in an atom. The direction towards the electron can be completely described by giving two angles. This is true on any sphere, for example the surface of Earth (where the position angles are called longitude and latitude), or in the sky (where directions are fixed by the angles of right ascension and declination). Thus, quantum angular momentum has only *two* independent components. In classical mechanics, of course, we are accustomed to angular momentum as a vector with *three* components! But the intrinsic uncertainty of the quantum world allows us only two; if these are known, the third is indeterminate. We can let expediency decide how to choose these two numbers, but unless there are reasons to do otherwise, we choose the total angular momentum J and the component J_z of the angular momentum along a fixed but otherwise arbitrary axis.

The second consequence of the uncertainty relation for angles and angular momentum is that if $J = 0$, then $\Delta\phi = \infty$. If we know with certainty that a particular state, e.g. an electron in an atom, has no angular momentum, then the direction to that electron is completely indeterminate. Therefore, *a state of zero angular momentum is spherically symmetric* (but not necessarily a uniform sphere; there can be radial structure, as we will see presently).

An electron with $J = 0$ in an atom is seen in no direction in particular; this shows how frightfully wrong it is to think of an atomic electron as orbiting! It also shows just how weird quantum mechanics seems to us, with our daily-life intuition. We can best visualize how a single electron can occupy a spherical volume by thinking strictly in terms of observation: if I were to measure the position of a $J = 0$ electron a million times, I would obtain a cloud of position points that would fill a spherically symmetric volume in space. Such a spherical $J = 0$ state is called an S-state in quantum mechanical jargon; however, the indication S comes from an ancient classification of certain spectral lines that were deemed to be special for being *S*harp. By coincidence, it turns out that the electron states of atoms that are responsible for these spectral lines are *S*pherical.

7.5 Three-dimensional rotations

Because space is symmetric under rotations about any arbitrary direction, we must not only consider rotations in a plane, but also those of three-dimensional space. Clearly, rotation in a plane is a special case of rotation in space; plane rotations are a subgroup of the rotations in space. Because space has three dimensions, spatial rotations can be composed from rotations about three different (e.g. mutually perpendicular) axes.

A phenomenal example of a symmetry group in action, based on the rotation group in three dimensions, is Rubik's Cube. Those of you who

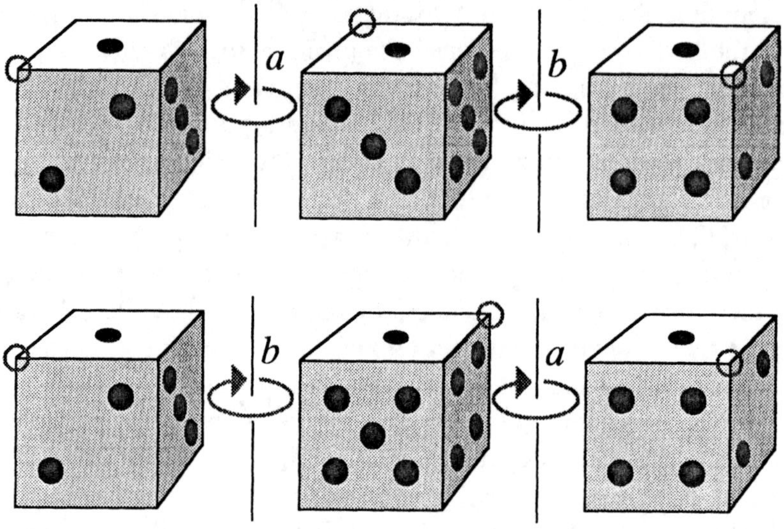

Fig. 7.7 Successive rotations *a* and *b* of a die about a fixed axis. The result of the rotation can be traced by following the corner marked with a grey circle. By comparing the dice on the right, we see that the result of such rotations does not depend on the order in which the rotations are executed: for rotations about a single axis, $a \otimes b = b \otimes a$. These rotations are said to 'commute'.

have mastered it will know what I mean; those who have not are referred to the book by Frey and Singmaster. Here we will consider a much simpler example, that is yet tricky enough: an ordinary die.†

Get yourself a pair of dice. While playing around with them, you will notice two *very* important things: first, that the *order* in which rotations are executed can make a difference; second, that two successive rotations about different axes can always be replaced by a single rotation, but the axis of that rotation points in an unexpected direction. We will see later in the story what these things mean when groups are applied to particles.

Place a die before you, the 1 on top, the 2 facing towards you (Fig. 7.7). Then the 3 is on the right. We count turns, as seen from above, clockwise. Hence a 90° turn about the (1–6) axis brings the 3 up front and the 5 to the right. Now start again, having the 2 facing towards you; first execute a 90° turn about the (1–6) axis, and then a 180° turn likewise. This should

† The notation given in the text will only work for correct dice. A correct die is cubical, with one through six dots engraved on the faces, distributed in such a way that the total number of dots on opposing faces is seven, and such that there is one corner at which the one-, two- and three-dot faces join in counterclockwise direction.

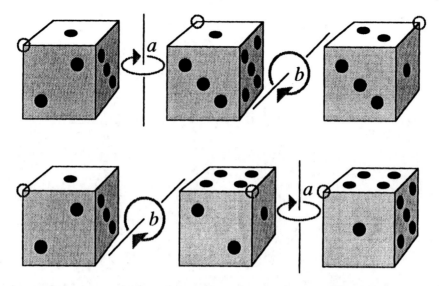

Fig. 7.8 Successive rotations a and b of a die about two different axes. The result of the rotation can be traced by following the corner marked with a grey circle. By comparing the dice on the right, we see that the result of such rotations *does* depend on the order in which the rotations are executed: for rotations about a single axis, $a \otimes b$ does not equal $b \otimes a$. These rotations are non-Abelian (non-commutative): 'first-a-then-b' produces a different orientation than 'first-b-then-a'.

bring the 2 to the right and the 4 facing you. Do the same with the other die, starting in the same position, but turning 180° first and then 90°. It is apparent that, in this special case, the order of moves makes no difference. Or, in other words, $a \otimes b = b \otimes a$. A group for which this holds is called commutative or Abelian. For example, the group of all rotations of the die about one axis is Abelian.

Next, place the die in the same starting position (Fig. 7.8). First execute a 90° turn about the vertical (1–6) axis. This brings the 3-face front, as before. Second, execute a 90° turn about the horizontal (3–4) axis pointing away from you, that is, clockwise with the 3-face towards you. This brings the 2-face up. Now do the same with the other die, but with the moves reversed. First, a 90° turn about the horizontal axis (which now passes through the (2–5) faces!), bringing the 4 on top. Second, 90° about the vertical axis, passing now through (4–3): this leaves the 4 on top, and brings the 1 up front! Clearly, the order of moves does make a difference. Hence we call such a group non-commutative or non-Abelian. Apparently, a non-Abelian group has a kind of memory of the order of

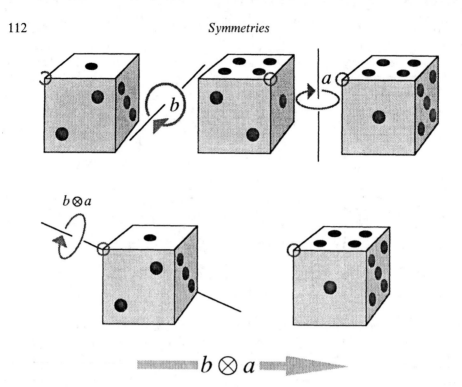

Fig. 7.9 Addition of rotations: the axis of the rotation 'first-*b*-then-*a*' does not relate in an obvious way to the axes of the separate rotations *b* and *a*. Instead of rotating about an axis through a face or an edge, the rotation axis of the composite operation $b \otimes a$ goes through a corner.

events; this provides an extra richness that will turn out to have spectac-ular consequences when we try to interpret the behaviour of elementary particles as due to a non-Abelian symmetry (it also makes Rubik's Cube so difficult).

Let us call the operation that rotates the die over 90° about the (1–6) axis *a*, and a 90° turn about the (3–4) axis *b*. We just saw that $a \otimes b$ does not equal $b \otimes a$. But because the rotations form a group, we know that there must be a single rotation *c* such that $a \otimes b = c$. Likewise, there is a single rotation *d* such that $b \otimes a = d$. Now we ask: with the above definitions for *a* and *b*, what is $c = a \otimes b$? Try it, and you will be surprised (Fig. 7.9). The combined rotation *c* is a turn over a third of a circle, 120°, about an axis through the *corners* of the cube, namely the corner where the (1–2–3) faces meet and those where (6–5–4) meet!

Thus, we do *not* get a simple addition of rotation angles; in some sense, we have '90°+90° = 120°', a very peculiar result. And the combined rotation axis does not lie somewhere midway between the axes of *a* and *b*, but jumps out in a different direction in space! Find out for yourself what $d = b \otimes a$ is. This amazing group behaviour is intimately related to important particle properties, to which I will return in Chapter 10.

8

Quantum relativity: nothing is relative

☆

8.1 Rotations in space-time

So far, I have discussed in detail two of the ingredients that govern the structure of our Universe: quantization and symmetry. Now we will consider the third, namely relativity. As it happens, relativity is actually associated with a symmetry, albeit a rather unusual one. Whether this is a coincidence or whether it indicates a deeper level of physical truth is not understood at present. As an introduction to the symmetry that produces relativity, let us return for a moment to rotations in space.

A symmetry group always leaves something unchanged; in the case of space rotations, the invariant is the distance between points. In three-dimensional space, we need three numbers (coordinates; say x, y and z) to describe the position of a point. The distance D is given by the Pythagoras recipe: $D^2 = x^2 + y^2 + z^2$. In a plane, two numbers (say x and y) suffice. Under rotations of the plane, $D^2 = x^2 + y^2$ is an invariant.

Suppose that some sixteenth-century navigators in Amsterdam were to measure the latitudes and longitudes of two points on Earth, and from these calculate the distance between the points. Now let some Parisian navigators do the same. In olden times, people were every bit as chauvinistic about the location of their country as they are today, so naturally the Dutch used the meridian of Amsterdam, and the French used the Paris meridian for their observations. Consequently, the coordinates of the points they measured were different, but the distances between points – calculated using the Pythagoras recipe – were the same! If they had asked some British navigators for additional observations, they would have found yet another batch of coordinates (using the Greenwich meridian, of course!) but the distances would again come out the same. Some clever geometer might then have concluded, purely from the fact that the Pythagoras distance

$D^2 = x^2 + y^2 + z^2$ is invariant, that *Earth is symmetric under rotations*. If an even brighter geometer had extended this to the notion that space itself is similarly symmetric, a lot of physically useful things would have followed, for example the conservation of angular momentum.

Now let us reconsider what was said in Chapter 1 about the speed of light. The experiments done by Michelson and Morley showed that the speed c with which light propagates is independent of the way in which the source or the observer moves. As will be shown presently, this implies that there is some sort of superdistance in space and in time which is invariant under a corresponding superrotation.

The position x at time t of a signal that moves with speed c is given by the usual equation: (distance) = (speed)×(time), in symbolic form $x = ct$. Or, if you move backwards, $x = -ct$. In order to account for both possibilities, we will write $x^2 = c^2t^2$, or alternatively $x^2 - c^2t^2 = 0$. In the Michelson–Morley experiment, it was found that the speed c is always the same. From this, we can conclude that the quantity $x^2 - c^2t^2$ is an invariant: it must always be the same, no matter what the location x and time t of the observer are, and no matter how that location changes in the course of time. Apparently, there is some kind of superdistance which we can use in space and in time, calculated by means of the `Minkowski recipe`: $S^2 = x^2 - c^2t^2$. And, just as the invariance of D^2 in space forces us to conclude that space is symmetric under rotations, the invariance of S^2 in space and in time clearly suggests that space and time (henceforth called `space-time`) are symmetric under some sort of superrotation.

8.2 The light cone, our compass in space-time

The invariance of c has the bizarre and dramatic consequence that the measurements of intervals in space and in time are no longer independent. This can be seen intuitively by noting that it is a *velocity* which must be constant, and a velocity has a mixed dimension: $[c]$ = (metres per second) = (space)/(time). In order to keep the speed c invariant, *space and time must conspire*: the yardsticks in space and in time must be interrelated so that measurement of space and time intervals always produces the same value of c.

This is shown most clearly by the `Lorentz light clock`. This hypothetical instrument (Fig. 8.1) consists of two parallel flat mirrors, between which a light ray bounces back and forth. Each round trip of the light ray corresponds to the tick-tock of the clock; let this time interval be $2t$. If the distance between the mirrors is D, then we have $t = D/c$. That is to say, the

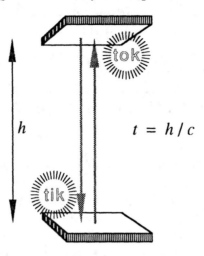

h

$t = h/c$

Fig. 8.1 Stationary Lorentz light clock: a light ray caught between two flat parallel mirrors, a distance h apart. Time t is measured by the transit time h/c between the mirrors. Because the speed c of light is always the same, such a clock is a perfect timepiece.

tick and the tock each take t seconds when the clock is at rest. Why is the clock rate different when the clock moves? *Because c must be kept the same.*

Let the clock move with speed v in the direction of the plane of the mirrors (Fig. 8.2). The light ray crosses from one mirror to the other, but now it shoots at a moving target. The target mirror moves a distance $x = vT$ during the tick interval, where T is the tick-time kept by the moving clock (as opposed to t for a stationary one). Because the mirrors are a distance D apart, the light ray now must cover a path with length $L = \sqrt{x^2 + D^2}$, by Pythagoras's rule.

Now comes the crunch: the distance L is traversed *with the same speed c* with which the distance D was crossed! Thus, we have a moving clock in which each tick covers $T = L/c$ seconds. Using Pythagoras's rule for L, we obtain $T^2 = (L/c)^2 = (x^2 + D^2)/c^2$. Or, if we work it all out using the fact that $x = vT$ and $t = D/c$, $T = t/\sqrt{1 - v^2/c^2}$.

Sorry for the algebra, but I just *had* to show you explicitly how simple it is, and how immensely odd: a clock that ticks t seconds when at rest, ticks a factor $\sqrt{1 - v^2/c^2}$ *slower* when seen by an observer who moves with speed v with respect to that clock! This factor, called the Lorentz factor (usually called γ) can be enormous when v approaches c. But in our daily life it is insignificant: because c equals 300 000 kilometres per second, we

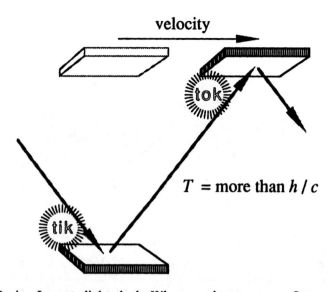

velocity

tok

T = more than h/c

tik

Fig. 8.2 Moving Lorentz light clock. When an observer sees a Lorentz clock move, the light ray between the mirrors must follow a slanted line, which is longer than the perpendicular distance between the mirrors. Because the speed of light is the same for all observers, the clock is seen to tick more slowly: it takes more time to traverse the slanted line than the perpendicular distance.

see immediately that even the fastest human spacecraft, moving at about 30 km/s, has a Lorentz factor which differs only a few *billionth* from unity.

The rate of a clock depends on the way it moves with respect to the observer. This is called `time dilatation`. An argument similar to the one with the light clock shows that how long something is seen to be depends on its relative motion, too. This is called `Lorentz-FitzGerald contraction`. Time and space measurements are no longer independent or absolute: such measurements are not meaningful without reference to one's relative state of motion. But what yardstick can we then use to find our way in the world? Clearly, we cannot obtain meaningful results if various observers see the same thing in different ways. Well, the bad news is that space- or time measurements themselves are no longer appropriate. The good news is that *the speed of light itself can be our yardstick*. What better ruler to use than one which never varies? So let's use the invariance of c to our advantage!

Thus, the propagation of light through space and time can serve us as a guide for all measurements. The fun of this is that it is very graphical. All

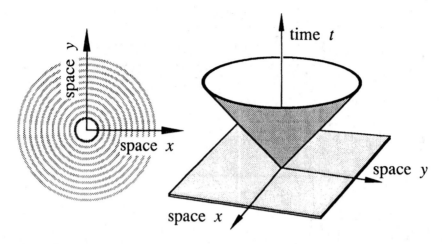

Fig. 8.3 The light cone in two dimensions built up from a stack of ripples with increasing diameter. We can retrieve the size of the expanding light wave by cutting the light cone with a plane perpendicular to the time axis. In reality, the cone is four-dimensional and is built out of a stack of spheres instead of circles.

we have to do is to trace a flash of light (Fig. 8.3). If we set off a flash at time zero, and let it move for t seconds, it will have spread out in a sphere with radius $r = ct$, just as a stone thrown into water causes a circular ripple to spread outwards. Suppose that we make a film of the spreading ripple, and stack the individual frames above one another. Then we see that the ripple traces a surface in space-time. This surface is a cone when the ripple is two-dimensional. In the case of light, the ripple is an expanding sphere, and we obtain a hypercone in four-dimensional space-time. Because that's a big mouthful, and because it's easier to draw in three dimensions, we usually speak of the *light cone*.

What's special about the light cone? Because the speed of light c is invariant, *the apex angle of the cone is the same* under all circumstances. Because c never varies, the shape of the light cone never varies either. Thus, the light cone can serve as an invariant object, a compass for navigating in space-time (we will use this in Chapter 10 when discussing gravity).

8.3 The Lorentz transformation

The superrotation that leaves the Minkowski distance $x^2 - c^2 t^2$ (or, in general, $x^2 + y^2 + z^2 - c^2 t^2$) unchanged, is called the `Lorentz transformation`.

It is completely analogous to rotations in ordinary space, although the superdistance in space-time contains a minus sign: $x^2 - c^2 t^2$.

The novel and unexpected feature of the Lorentz transformation is that Lorentz symmetry is a property of space and time together. It is precisely this feature which leads to the counter-intuitive results of relativity physics. Therefore, it is worth while tracing the ancestry of this curious mixing of space and time. It all comes from the fact that it is a *speed* that is constant: in the expression of speed, space and time are necessary ingredients, because (speed) = (distance)/(time). Thus, we cannot escape the necessity of using space as well as time in the expression for the invariant. Recalling Einstein's question 'How does a light ray see another one move?' we note that in order to keep c invariant, speeds must add in unexpected ways. To make this possible, our classical accounting for distances in space and time must be modified.

If we restrict ourselves to motion in a straight line, we have $x = \pm ct$ for light rays and $x = vt$ for things that move more slowly than light. The corresponding invariant is $x^2 - c^2 t^2$, and the Lorentz transformation follows (the restriction to the special case of straight-line motion is commemorated in calling the theory *special* relativity; the restriction can be removed, as we will see in Chapter 10).

In space the collection of all points with a fixed distance to a given point is a sphere ($x^2 + y^2 + z^2 = D^2$ = constant); in space-time this locus is a hyperboloid ($x^2 + y^2 + z^2 - c^2 t^2 = -S^2$ = constant). The invariance of the speed of light implies that the correct measure of distance in space-time contains a *minus sign*. This imposes a very peculiar structure (Fig. 8.4), namely one in which space-time is divided into three parts: points with a `space-like` distance, points with a `time-like` distance and points on the `light cone`.

For simplicity, let us collapse all three spatial dimensions into one, called x; then the Lorentz 'circle' obeys $x^2 - c^2 t^2 = -S^2$. Suppose that $-S^2$ is positive; all Lorentz circles with this property occupy a certain finite region in space-time. This region is the collection of all points that have a space-like distance to the origin $x = 0$, $t = 0$. Contrariwise, the Lorentz circles with a negative value of $-S^2$ comprise all points with a time-like distance to the origin (0,0). The reason for these names is straightforward: travel between points that have a time-like separation is possible with a speed less than the speed of light (all it takes is time to cross that distance). Space-like points, however, are truly separate, in the sense that one cannot go from one to the other without moving faster than light. The points with $S^2 = 0$ are traced by $x = \pm ct$, i.e. lines along which light travels in space-time. Thus, $S = 0$

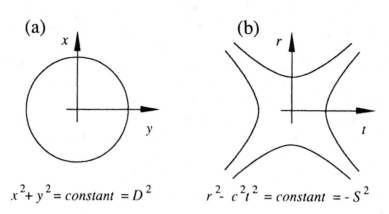

$$x^2 + y^2 = constant = D^2 \qquad r^2 - c^2t^2 = constant = -S^2$$

Fig. 8.4 Circles in space and in space-time. A circle is the locus of all points with a fixed distance from a given point. (a) A two-dimensional circle with the Pythagoras distance recipe $x^2 + y^2 = D^2$. (b) A circle with the Lorentz distance recipe $r^2 - c^2t^2 = -S^2$.

defines the light cone (in one dimension, this cone degenerates into a pair of lines). The time-like points can be said to lie inside the light cone, whereas the space-like points lie outside it.

8.4 Antimatter

The time-like and space-like regions differ in a characteristic and peculiar way when we consider what is meant by 'simultaneous'. In classical mechanics, there are no difficulties with this concept, but the peculiar mixing of space and time in relativity plays some funny tricks. It turns out that in the time-like region, the order in time of events is always the same, even for different observers. Suppose that I have two helpers, say Laurel and Hardy, and that I see Laurel doff his hat before Hardy (Fig. 8.5). An observer who has a time-like distance with respect to me will always agree: Laurel saluted before Hardy.

However, an observer who, with respect to me, has a *space-like* distance will not always agree. Depending on that observer's velocity with respect to me, the time order of events may be *reversed*: Hardy may be seen to take off his hat before Laurel!

The point about relative time ordering deserves some extra attention, because it is associated with a very remarkable prediction of quantum

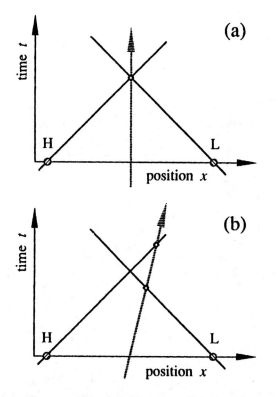

Fig. 8.5 Space-time diagram showing how a stationary observer sees Laurel and Hardy take off their hats (a), and how a moving observer sees the same events (b). Events which a stationary observer sees as simultaneous happen at different times as seen by a moving observer.

relativity: the existence of antimatter. Let us set up an experiment to check whether events that appear to be simultaneous for one observer also appear to be simultaneous as seen by others. To Laurel and Hardy you give watches which are identical to your watch and which have been synchronized with yours. You ask Laurel to walk away from you for 1 kilometre, wait until his watch indicates noon, and at that moment to take off his hat. You ask Hardy to do the same, except that he is to walk away in the direction opposite to Laurel. Since you are standing exactly between them, you naturally expect to see them taking their hats off simultaneously. Now you repeat the experiment just as before: you are again precisely midway between Laurel and Hardy at the crack of noon, but now you arrange things so that, at that precise moment, you are moving in a straight line towards Laurel. In this second experiment, you will

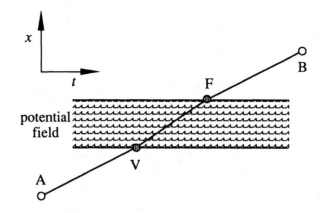

Fig. 8.6 Electron moving through a plane-parallel field, showing the space-time path of an electron that is accelerated at V and decelerated at F by a potential field (zone with wavy lines).

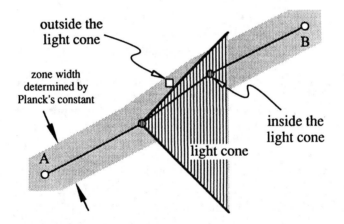

Fig. 8.7 Quantum spillover outside the light cone. Here we see the same electron path as in Fig. 8.6, but now it is a quantum path that is fuzzy due to Heisenberg uncertainty. Thus, some points of the path lie in the space-like region outside the light cone, as indicated by the white square.

observe Laurel taking his hat off *before* Hardy! The connection between the first and the second experiment is given by the Lorentz transformation.

Thus, because of the mixing of space and time, different observers no longer necessarily agree on the time order of events! As we will presently

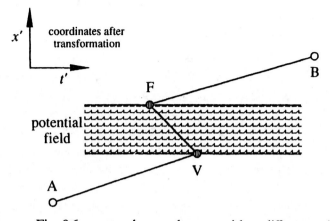

Fig. 8.8 Same as Fig. 8.6, as seen by an observer with a different velocity (i.e. as seen after a Lorentz transformation). Notice that the order of the events V and F in time is now reversed!

show, this has astonishing consequences: it leads literally to the appearance of a whole new world (Fig. 8.6). Let us take a particle, say an electron, at a certain initial point A, and propel it through an electric field towards a point B. At V, the electron is speeded up by the action of the field; at F, it leaves the field, slows down again, and proceeds to B. As drawn in this diagram, event V precedes event F. Verify this by scanning the picture with your Feynman diagram scanner.

Since the electron cannot move faster than light, its path from V to F must lie within the light cone: V and F have a time-like separation. Thus, V would precede F as seen by other observers, too. This holds strictly in classical relativity, but when we take quantum behaviour into account as well (Fig. 8.7), we find that *the space-like region is no longer a no-man's-land.* The reason for this is that, due to Heisenberg's uncertainty relations, a wave packet is not exactly pinpointed. Thus, the woolly path of a quantum spills over just a little from the time-like region into the classically forbidden space-like zone.†

But once a quantum at point F has set foot in the space-like region with respect to point V, the time order of events V and F is lost (Fig. 8.8). Accordingly, there must be observers *for whom* F *precedes* V! What are we to make of this? We could say, of course, that the electron runs backward in time from V to F, but that seems somehow repugnant. Part

† Actually, this is only strictly true for quanta with positive energy; see Feynman's essay *The Reason for Antiparticles*, the first of two articles in *Elementary Particles and the Laws of Physics*, by R.P. Feynman and S. Weinberg.

of the unpleasantness is that a particle which runs backward in time has a negative frequency; and this, by De Broglie's rules, means a negative energy – a most unpalatable prospect. However, there is another interpretation which suggests itself immediately when you scan this picture with your Feynman diagram scanner. The electron leaves A; shortly thereafter, a pair of particles spontaneously appears at F; one of these particles proceeds towards the original electron, with which it merges such that both disappear at V; meanwhile, the other particle proceeds from F to B.

But wait – something looks wrong here. At first glance, considering the created particle as possibly real seems senseless, because we would be obliged to give it a positive energy, and where would that come from? The demand that all energies be positive would be incompatible with relativity and classical mechanics. Surely, pair creation and annihilation violate the conservation of mass and energy! Can we give up this conservation? Yes we can, at least temporarily. The uncertainty relations allow us to embezzle a bit of mass, as long as it is given back promptly. To this must be added the fact that the quantum uncertainty relation $\Delta E \, \Delta t > \hbar$ by itself does not allow the temporary creation of a particle; but the relativistic relationship $E = mc^2$ comes to the rescue, and we can write $\Delta m \, \Delta t > \hbar/c^2$. Thus, only if relativity and quantization are *combined* do we obtain a description in which massive antiparticles appear.

This is truly breathtaking: because of the invariance of the speed of light, space and time must mix so that the time order of events in space-time is not immutable. Furthermore, quantum uncertainty allows a particle to spill over just a little into space-like no-man's-land. Because of these two facts, we find that to each particle there corresponds an antiparticle. A particle and its antiparticle can appear spontaneously at a space-time event; this is called `pair creation` (e.g. point F). Similarly, they can merge and disappear in an `annihilation` (point V). The doubling back of the particle's path means that particles must be created or annihilated in pairs, and not in threesomes or whatever. Incidentally, it is common practice to indicate an antiparticle by placing a bar over the symbol of the corresponding particle, e.g. v is a neutrino and $\bar{v}$ an antineutrino; p is a proton and $\bar{p}$ is an antiproton. Some particles are their own antiparticles; for example, the photon γ is identical to the antiphoton $\bar{\gamma}$.

The joining of relativity and quantization allows us to make the notion of smallness more precise. As we have seen, the De Broglie relationship defines the length scale $\lambda = 2\pi\hbar/mv$. If the speed v can increase without bounds, we could have arbitrarily small length scales, in which case you might wonder why I made such a fuss about an absolute scale of smallness. However, v

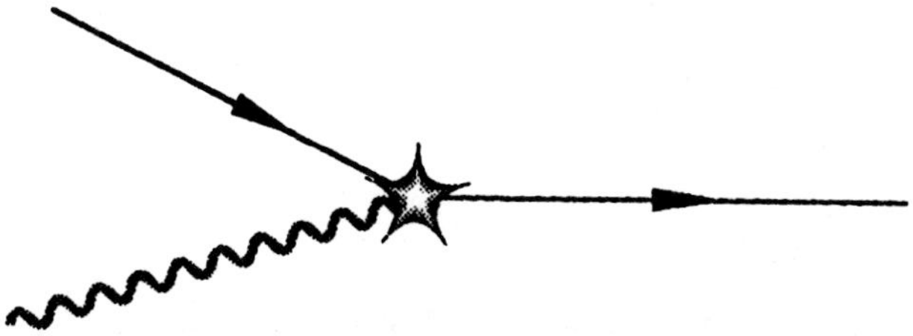

Fig. 8.9 Electron–photon vertex. Diagrams of this type are the basis for all quantum interactions: they contain an incoming fermion, an outgoing fermion and a boson, which couple at the vertex.

cannot exceed the speed of light c; therefore, with each mass m is associated a smallest length scale, called the Compton length $\lambda_C = \hbar/mc$ (whether or not the factor 2π is carried over from the De Broglie expression is a matter of definition). The Compton length is something like the size of a particle with mass m.

That the Compton length indicates the extent to which a particle is smeared out in space can be seen as follows. According to the uncertainty relation for energy and time, the best possible combination of energy and time uncertainty is $\Delta E \, \Delta t = \hbar$. Because of relativity, ΔE cannot be less than mc^2, so that $m \, \Delta t = \hbar/c^2$. If we call λ_C the size of mass m, then the particle cannot be located in time to a better precision than $\Delta t = \lambda_C/c$, because one cannot go from one side of the particle to the other faster than with speed c. Thus we find $\lambda_C = \hbar/mc$, as before. Notice again how quantization and relativity are intertwined: from $\hbar$ and m alone we cannot build a quantity with the physical dimension of a length, nor can we build one from c and m; but from $m, \hbar$ and c we can make exactly one length scale, namely λ_C.

The existence of antiparticles, pair creation and annihilation must be built into† any description of the forces in our world, because it so happens that the space-time of which our Universe is built is Lorentz symmetric. We would be wise to construct a picture of these forces in which that Lorentz symmetry is manifest. Thus, building the interaction between a particle and a field quantum out of separate building blocks like those shown in Fig. 8.9 would be unwise, because such a building block is individually not Lorentz invariant. After all, we have just seen that the different

† From now on, I will assume as a matter of course that antiparticles are included when we discuss 'fundamental' particles. For example, when I introduce a colour triplet of quarks, I will not state explicitly that there is a corresponding triplet of anticolour antiquarks.

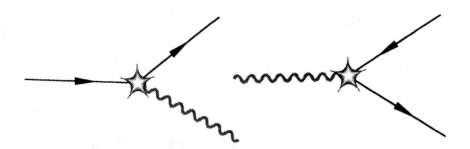

Fig. 8.10 Electron–photon vertex as in Fig. 8.9, with a different ordering of ingoing and outgoing states.

time ordering of events can make the very same interaction between field and particle look like Fig. 8.10 simply by dragging the quantum paths around.

Accordingly, we should not split up the interaction into such puny pieces, but instead use a global approach, dealing only with overall exchanges such as the one shown in Fig. 8.11, in which it is understood that the time order of the vertices V and F is irrelevant.

Another compelling reason to take a global point of view is this: the possibility of pair creation means that *the total number of particles in an interaction is not constant* – now there's a challenge for relativistic quantum billiards players! Not only must we use vaguely shimmering balls, floppy cues, and cushions through which the balls can sometimes pass without jumping over, but also the number of balls is potentially infinite. How can a decent theory ever be built from such ingredients?

8.5 Taking the large view

Well – it proved nearly impossible to build that theory; nearly, but not quite. As it happens, the situation can be saved by accepting the overall picture of quantum space-time as one whole, and not needlessly chopping it up into pieces that are individually unmanageable. It is one of the singular beauties of the Feynman diagrams that they provide precisely this global view. Thus, they are much more than pictographic bookkeeping. The Feynman approach keeps the building blocks of a force manifestly Lorentz symmetric. Of course it is possible to build a house from complex clumps of clay, but how much easier it is to use symmetrical bricks!

Moreover, the global approach inherent in the Feynman diagrams capitalizes on the fact that, on the submicroscopic scale of quantum particles, we

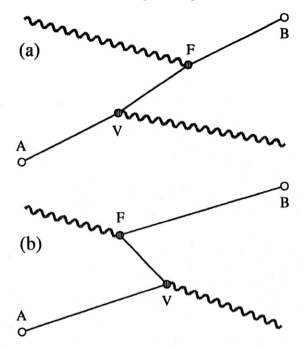

Fig. 8.11 Quantized version of the electron motion from A to B shown in Fig. 8.6 and Fig. 8.8. The interaction with the potential field has been replaced by the interaction with velpons (field bosons) at V and F.

are dealing with intrinsically small objects. When we deal with interacting particles, we must necessarily descend into the realm of the very small, where there is an intimate exchange of quanta, and where it is not possible to deduce who absorbed or emitted what and when. In a Feynman diagram, the question of the interaction between particles is not chopped up into emission and absorption of a field quantum as separate events, but is considered in a holistic way, as a direct interaction, the *exchange* of a quantum: precisely the kind of direct contact that Descartes would have found amusing.

Not only does the global view represented by the Feynman diagrams acknowledge the intimacy of quantum interactions, but it also eliminates problems that would arise from the fact that the speed of light cannot be exceeded. Because of the *c* limit, instantaneous interactions as proposed by Newton are impossible; therefore, *it is not enough to prescribe the present state of a system if we want to know how it will behave in the future.* Because *c* is finite and maximal, all interactions are underway for a while; thus, we must know the entire status of all fields everywhere in the Universe in order to calculate the future from the present.

This is most awkward, as becomes apparent when we recall what happened to the electron traversing the electric field sketched above. For if we wilfully ignore the global view, we must wonder how the dickens the vacuum 'knows' that it is to produce a particle–antiparticle pair at event F, so that the antiparticle can annihilate the particle at V. This 'U.S. Cavalry model', in which an antiparticle magically arrives to annihilate a particle in the nick of time, may be traditional in spaghetti-Western films, but in quantum relativity it is clearly unsatisfactory. The speed-of-light limit forces us to treat space-time and its residents as a whole, and the Feynman diagrams represent just such a holistic view. They weave a web in space-time without having to bother with irrelevant stuff such as the time order of the web's nodes, or the number of intermediate strands of silk.

We can use the words small and large in a meaningful way because of the existence of the uncertainty rules of quantum mechanics. We can use slow or fast because of the existence of an absolute speed, the speed of light, and the associated laws of relativity. Thus, we see that in our Universe we do *not* find that 'everything is relative'. On the contrary, if we insist on making pronouncements like that, we have more reason to say that *everything is absolute*. There is an absolute size, the Compton length; there is an absolute speed, the speed of light. The combination of these absolutes, together with the laws of symmetry, produces all the richness of structure in Nature. So please stop muttering that all things are 'relative'. If it's any consolation to you, Einstein himself was unhappy with this term, which was not invented by him.

8.6 The dizzy world of spin

By now, you are probably convinced that the world of the fast and the small is utterly bizarre. It is difficult to imagine what it is like to be an electron that collides with another, if you are only accustomed to the things that happen when billiard balls collide.

Even more beyond the reach of intuition are the things that occur when particles rotate. These happenings are the stuff of the second half of this chapter. But just in case you think that the bizarre effects of spin are entirely restricted to the quantum world, let me show you just how peculiar rotating things can be. The point of this little excursion is to show you how a spinning object can respond by moving *perpendicular to a force* acting on it.

To prepare you for the amazing things that spinning quantum particles do, I am going to begin by showing some effects in classical mechanics that are pretty weird already. Star actor of this section: a simple top. Please beg, steal or borrow one, and proceed.

Place the top with its point on the table, and let go. Surely, it will fall sideways: because the support of the point is never exactly below the centre of the top, the pull of gravity increases the tilt of the top until it falls over (Fig. 8.12).

Simple, wouldn't you say? You have seen this a thousand times before. But now do the same with a top that is spinning fast. Instead of falling over, the top starts a slow rotation or wobbling, leaning away from the vertical at a fixed angle. Only after a long time, when friction has slowed the top down, does it lean over more and more, until finally it falls over on its side. This wobbling motion is called `precession` (Fig. 8.13). You can make the experiment even more dramatic by means of a top suspended in a metal ring (it is then called a `gyroscope`). The ring allows you to put the top with its tip at the edge of a table, with the spinning top hanging sideways in space, and yet – as long as it spins fast – *it does not fall*!

Bizarre, wouldn't you say? How can it possibly be that a rotating top does not fall sideways? Is rotation some sort of antigravity, or what? I am convinced that you have seen precession thousands of times, but how often have you wondered what trick the spinning plays on gravity? The amazing thing is that a spinning top on which a force acts does not begin to move in the direction of the force, but instead it moves *perpendicular to the force*. We will see later that this weird behaviour has fascinating quantum analogues, such as the *Lorentz force* acting on an electric current in a magnetic field.

Before you run off to the Patent Office to claim the invention of antigravity, let me explain how precession works. Actually, the explanation is a bit clumsy without vector calculus, but it will have to do; besides, without all the mathematics it is more transparent. There is no magic or antigravity involved: just good old-fashioned matter in classical motion. But, you will have to admit, rather different from what your intuition might expect of moving things.

Three ingredients play a role: first, the inertia of the matter of which the top is made, and the fact that its particles stick together to form a solid body; second, the rotation of the top; third, the pull of gravity (by the way, other forces would serve as well, but are much more difficult to experiment with).

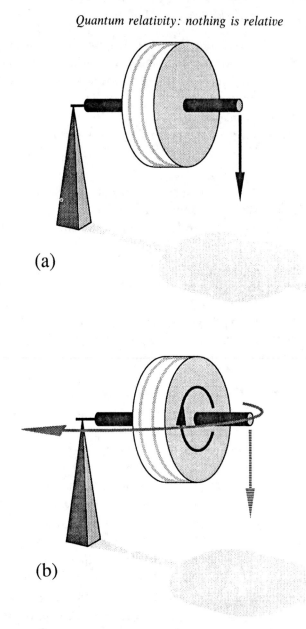

(a)

(b)

Fig. 8.12 (a) A top that does not spin, when supported only at its tip, will fall in the direction of the pull of gravity (solid arrow). (b) A spinning top, when supported at its tip, does not fall. Instead, it starts to move in a direction (dark grey arc) which is perpendicular to the pull of gravity (striped arrow).

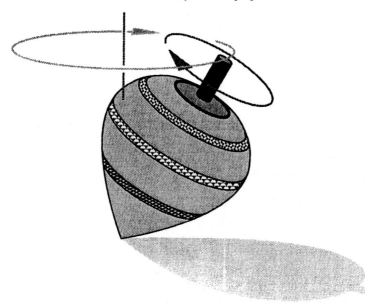

Fig. 8.13 A precessing top. Even though the top leans away from the vertical, it does not fall over as long as it spins (dark grey arc). Instead, it gyrates about the vertical (striped line), a motion called precession (light grey arc).

The first item means that the particles persist in their motion unless a force acts on them. The second means that these particles move in circles† around the spinning axis. The third ingredient causes the particles to deviate from their circular motion.

As we have seen in all previous cases, the actual motion of the particles of the top is composed of inertia-plus-force, moving-and-falling. This holds for the orbital motion of the planets as well as for the motion of our top; in our case, we have (combined orbit) = (rotation plus falling). For simplicity, let us place the top perpendicular to the force of gravity, and support it at one tip, just as in the case of a gyroscope placed sideways on the edge of a table (Fig. 8.14). Look straight along the axis of the top, and suppose that it is then seen to rotate clockwise.

If a small piece of the top were left to its own devices, it would continue with a fixed velocity in a straight line. Because it is constrained to move with the rest of the solid body of the top, it moves in a circle. The constraining

† Had I not said that a particle moves in straight lines, in the absence of a force? Yes, but there are hidden forces here, namely those that keep the matter of which the top is made together. They are in fact quantum forces, called Van der Waals forces, but it can be shown that these internal forces play no role in the argument here.

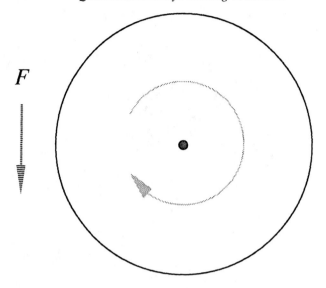

F

Fig. 8.14 First step in the explanation of precession. We are looking along the spin axis of the top; the pull of gravity F is directed along the striped arrow, perpendicular to the axis. The spindle on which the top spins is supported at the end which points away from us.

force is hidden in the material of which the top is made (it is known as a Van der Waals force, which we will encounter later).

Take a bit of matter that starts at the upper edge of the top. After a fraction of a second, in the absence of gravity, it would have rotated over a small arc. Likewise, in the absence of rotation, it would have fallen a small distance. When both effects are present, we obtain a combined rotating-and-falling motion. The combined orbit of a particle at the upper edge of the top is, as always, due to inertia-and-falling, in this case rotation-and-falling. Without gravity, the particle would have advanced along the black circle shown in Fig. 8.14. Without rotation, the particle would have fallen vertically. The combined motion carries the particle along the grey arc (Fig. 8.15).

Because the top is a solid body, this arc corresponds to the rim of the top after the small interval of time that has elapsed between the two black points on the arc. Rotation-and-falling means that the bit of the top must still move on a circle, albeit a different one from the circle on which it started.

Which one? We can find out in what direction the top has moved during our small time interval by passing a circle through the two black points (Fig. 8.16). Immediately, we see something quite surprising. The top has

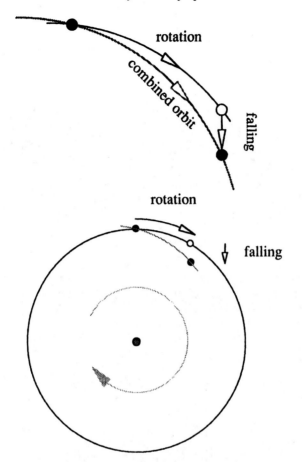

Fig. 8.15 The combined orbit of a particle at the upper edge of the top is due to rotation-and-falling. Without gravity, the particle would have advanced to the position of the open circle. Without rotation, the particle would have fallen vertically over the distance between the open and the closed circle. The combined motion carries the particle along the grey arc. Because the top is a solid body, this arc corresponds to the edge of the top after the small interval of time that has elapsed between the two black points on the arc.

indeed responded to the force of gravity, but *not* by moving in the direction of the pull! Instead, the axis of the top has moved sideways, *at right angles to the direction of the force*, and also, incidentally, at right angles to the axis of rotation.

This shift of the axis of rotation is called precession. Rotation is not antigravity, because the top has in fact done something, but it did not quite

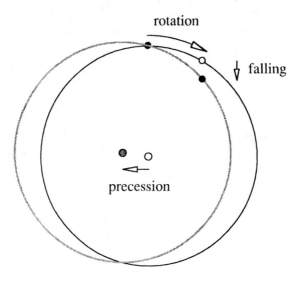

Fig. 8.16 The direction in which the top moves due to the force of gravity can be found by passing a circle through the two black points. In this way, we find that the disk of the top has moved from the position of the black circle to the gray circle. This is a shift at right angles to the direction of the force, and also at right angles to the axis of rotation. This is precession.

do what you might have expected. The force of gravity still has an effect, causing the top to 'fall sideways'.

An objection you might make is that perhaps this shift is due to the particular choice of the point of departure: the black point precisely at the upper edge of the top. To see whether that is true, let us take a point at the bottom edge (Fig. 8.17). Using the same construction, you can verify that there, too, the top is seen to shift to the left, by approximately the same amount. The fact that the shifted position of the axis does not quite coincide with the one found previously is due to the crudeness of the drawing. Instead of using motion during a vanishingly small time step (an infinitesimal amount of time), we must draw a finite arc between the black dots.

So you see that even classical mechanics can appear rather bizarre and non-intuitive, especially when we are dealing with spinning objects. The response of a spinning top to an external force can be really weird: instead of being accelerated in the direction of the force, the top can start to move sideways, at right angles to the direction of the force and to the direction of the pull of the force.

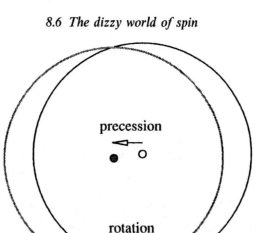

Fig. 8.17 The precession construction carried out for a point at the bottom of the top. Compare the shifted position of the axis with the one found in Fig. 8.16. You may find it interesting to do this construction for all points on the circumference of the top (things go faster when you cut a paper disk having the same size as the black circle, and use that to find the position of the grey circle). You will find that the axis always shifts to the left. Moreover, when you do the construction carefully you will see that the positions of the axis trace out a little circle that touches the original position of the axis. Under some circumstances, such a mini-wobble is in fact observed in moving tops; it is called `nutation`.

These very peculiar behaviours shows up in the quantum world as well. Explaining how is very convoluted, so I will just state them. First, an electric charge that moves in a magnetic field feels a force which is perpendicular to its velocity and to the direction of the magnetic field. This `Lorentz force` owes its surprising properties to the spin of the exchanged photon. Second, the exchange of a velpon can produce attractive as well as repulsive forces. The occurrence of repulsive as well as attractive behaviour in a single type of force is possible only if we take the spin of the exchanged quantum into account (here our beanbag exchange analogy goes rather wrong).

Apparently, spinning classical tops are funny beasts, so you may well expect that spinning quantum particles are even more bizarre. And indeed they are; so let us go back to the quantum world, and see what the properties of rotation become when the rotating bodies are absolutely small.

8.7 Spin and relativity

There is more richness yet in the amazing world of the very fast and the very small. The Lorentz clock told us that the existence of a universal speed (the speed of light) implies that time is relative: events cannot be given an absolute time ordering. We will now see that this has another peculiar consequence, namely the existence of *spin*, an internal degree of freedom which all particles possess. This property is called spin because it behaves like a rotation.

Infinitely small particles show the effects of spin, and the quantum rules impose a remarkable pattern on its behaviour. We will see below how essential this internal freedom of particles is for the action of the forces they produce. In fact, I have already anticipated this in the discussion about the apparent distinction between matter and force in our large-scale world.

To introduce the problems associated with spin angular momentum, let us consider the appearance of a spherical object flying in a straight line past an observer. Fig. 8.18 shows the flyby as it might look from a very large distance; next to it is the sphere as seen by the observer. We notice immediately that the sphere seems to rotate as it flies by, even though it does not rotate at all with respect to its own direction of motion. This apparent rotation is due purely to the motion of the sphere relative to the observer, wherefore it is called `orbital spin`. Notice that a second observer who moves with respect to the first will attribute a *different* orbital spin to the particle, because the path of the sphere relative to that second observer is different.

Now suppose that the sphere does rotate on an axis as it moves. Clearly, this rotation will add to the orbital spin, to produce a compound rotation. If we see such a particle fly by, can we separate these two effects? In classical mechanics we can, but in quantum relativity we cannot, because of the impossibility to define a time order of events. This is precisely the same phenomenon that gave rise to antiparticles: quantum uncertainty makes a particle stray into the no-man's-land of space-like distances, and in this region different observers can attribute a different time order to the same events. A spinning particle is like a clock, in which each turn indicates a tick; and because the timing of the same event differs for different observers, there can never be a universal rule that defines when the particle's spin clock strikes the hour.

Accordingly, the orbital spin and the rotation of the particle on its axis can never be disentangled. 'Where is a particle when' depends on the state of motion of the particle with respect to the observer: that's relativity.

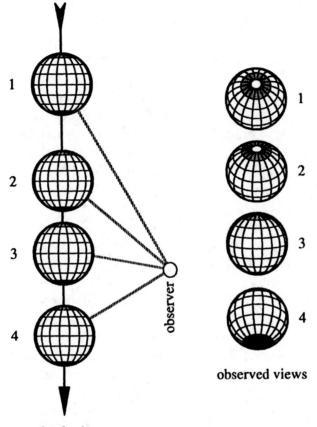

path of sphere

Fig. 8.18 A sphere that flies by an observer in a straight line is seen from different directions (labelled 1 through 4) as time goes by. The surface coordinate grid shows that this straight-line flyby is accompanied by an apparent rotation. This does not mean to suggest that the particle is actually a little sphere.

Therefore, any particle – even if it is not moving with respect to some observer – can behave as if it were rotating! The Lorentz symmetry of space-time, in conjunction with quantum uncertainty, endows each particle with a property, called spin, which behaves like a rotation.

Notice, incidentally, that the impossibility of separating orbital spin from rotation implies that it is meaningless to ask if the particle does 'really' rotate. *The spin of a particle is an internal degree of freedom, an overt consequence of the underlying Lorentz symmetry of quanta in space-time.* Spin is a relativistic residue, as it were, that remains even at speeds which are small compared with c (we encountered another such residue earlier in the famous $E = mc^2$).

The origin of spin is intimately related to the group behaviour of the Lorentz transformation, and therefore it is instructive to look into the origin of spin in a little more detail. We will start by using only Lorentz transformations in fixed directions, without any rotation at all. Such a rotation-free Lorentz transformation is called a `Lorentz boost`.

Basically, what happens is that two successive Lorentz boosts can *not* be replaced by a single Lorentz boost. Instead, we obtain a Lorentz boost *and a rotation*. Curiously, when we start without any rotation at all, rotation is thrown in for free. Lorentz symmetry means that rotation is unavoidable; that is the origin of spin.

Suppose that I look at the world as seen from Leiden. Someone zipping past me in the direction of Utrecht will see the same world, but Lorentz transformed; call the corresponding transformation U. Somebody else, moving towards Amsterdam, will see the world transformed by another Lorentz operation, say A. Now we can ask, what is the joint effect of U and A? We might expect that $U \otimes A = H$, where H is a Lorentz transformation in a direction intermediate between Utrecht and Amsterdam (say Hilversum). But that is *not* what we get. Instead, we obtain $U \otimes A = H \otimes (\text{rotation})$. The amount of rotation depends on the details of U and A, but the message is clear: the conjunction of relativistic motions spawns rotation. It is this property that thwarts attempts at distinguishing between spin and orbital angular momentum. We saw something quite similar in the combination of ordinary rotations (Fig. 7.9), and we will encounter this again in Chapter 14 (Fig. 14.7).

An attentive reader might remark that the above rough explanation of spin is unacceptable, because one might consider putting markers on the surface of the sphere that send out synchronized pulses of light. From the observed arrival times of these pulses, one could conceivably reconstruct the intrinsic rotation of the sphere. But this objection disappears if we make the particle small enough. Again we use the fact that small size has an absolute meaning (recall our discussion of the Compton length). When we descend into the realm of the very small, we enter the domain of quantum uncertainty, which does not allow us to distinguish which marker on the particle's surface flashed when. Thus, relativity and quantization combine to produce the degree of freedom we call spin.

8.8 The quantum of spin

Spin is quantized: it appears exclusively as a whole multiple of a fixed unit. We can see how this comes about by imagining a particle as it shrinks to zero

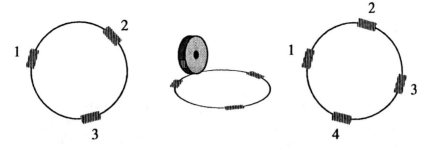

Fig. 8.19 Closed phase paths with $n = 3$ and $n = 4$ nodes. To construct this type of phase path we roll our muddy wheel along a closed track in such a way that the mudprints coincide after a whole number of turns around the track.

size. Consider a little rotating ball. Because of the quantum behaviour of Nature, there are wave properties associated with the ball; in particular, the wave has a certain phase. Now it is clear from our discussions of interference that the quantum path around the equator of the ball must be in phase with itself. Otherwise, it would interfere destructively with itself, and the particle could not be stable. Just as a guitar string can vibrate only in a fixed and ordered sequence of oscillations, a spinning quantum particle can rotate only in certain special ways; spin is quantized.

To see what the quantum of spin is, we must consider a closed quantum path that is locked in phase, so that the nodes (= the mudprints left by the wheel tracing the path) coincide after the path has wrapped itself around the spinning particle's equator (Fig. 8.20). Here we are beginning to stretch the metaphor a bit too far, but the mathematics is exact – and difficult! Clearly, this requires that the path closes after a whole number n of revolutions of the muddy wheel. If the wavelength laid down by the wheel is λ, and if R is the radius of the wheel's path, we obtain $2\pi R = n\lambda$. Now De Broglie's rule connects λ to the momentum mv by $\lambda = 2\pi\hbar/mv$, so that we obtain $mvR = n\hbar$.

We recall from Chapter 7 that mvR is the angular momentum; because we are dealing here with the rotation of the particle itself, we speak of spin angular momentum or simply spin, usually indicated by S. Thus we have $S = n\hbar$; notice the appearance of the number n, an instance of a quantum number indicating an internal degree of freedom of the particle. The number n is an integer: one can have $n = 0, 1, 2, \cdots$. By the way, we see that it does not matter that we have used a finite radius R for the path of the phase wheel; we have merely used it as an accounting device for the

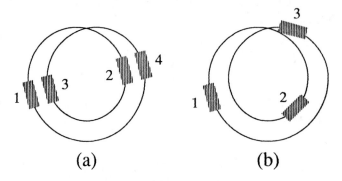

Fig. 8.20 (a) Double loop around the equator with $n = 4$, to produce spin $2\hbar$. (b) Double loop with $n = 3$, to produce spin $3\hbar/2$.

phase, which is a pure number that does not depend on physical size. Thus, the final expression for the quantization of S is independent of R.

What if the phase wheel loops around more than once? After all, we only require that the path closes after a whole number n of revolutions of the muddy wheel, which could equally well happen after having gone *twice* around the equator of the particle. First, let the number n be even (say $n = 4$). In that case, we see straight away that the third mudprint coincides with the first, and the fourth with the second. Thus, we obtain a superposition of two indistinguishable paths, each with $n/2$ nodes; but because n is even, $n/2 = m$ is a whole number, and we end up with a class of particles that have their spin quantized according to $S = m\hbar$, where m is an integer just as we had before.

Second, let n be an odd integer: $1, 3, 5, \cdots$. Then the mudprints do *not* coincide in pairs any more, but there is an *anti*coincidence. Each point on the second loop traced by the phase wheel is exactly midway between two points on the first loop. In that case, the two loops are not the same, and we obtain a particle with an equator of which the circumference $2\pi R$ equals $n/2$ times the De Broglie length, so that $2\pi R = n\lambda/2$. Thus, we obtain a class of particles that obey the spin quantization $S = n\hbar/2$, where n is an odd integer.

Further study of the phase process outlined above shows that more loopings of the phase wheel produce superpositions that can be reduced to the case of one or two loops. Thus, the two classes of particles obtained are the only two that occur. A particle from the first class, having a spin that is an integral number of Planck units ($S = m\hbar$), is called a Bose–Einstein particle or boson. One from the second class, where the spin is a half-integral number of Planck units ($S = n\hbar/2$), is called a `Fermi-Dirac particle` or `fermion`.

8.9 Indistinguishable processes

When we look at the phase-wrapping diagram of a boson, we see that it is a simple circle. Thus, we expect that a Bose–Einstein particle, turned a full circle of 360°, is indistinguishable from its original. Splendid: we have already noted that space is symmetric under rotations, so it is entirely expected that a full rotation turns a particle into itself. But when we try to do the same with a fermion, we notice something amusing: the phase-wrapping diagram consists of *two* loops, so that a fermion does *not* produce an exact replica of itself when rotated 360°! Instead, it requires two full turns, 720°, to make a fermion identical to what it was before the rotation.

This curious fact is related to a startling consequence, which becomes apparent when we ask: what state do we obtain when we rotate a fermion 360°? On the one hand, we have just argued that one full turn does not produce the exact same amplitude of the fermion. On the other hand, we have argued time and again that space is symmetric under rotations, which would seem to require that we must reproduce the original particle after a 360° rotation. The only way out appears to be that the amplitude of the original fermion is the precise opposite of the amplitude that is obtained when it is given a 360° turn: a single full rotation inverts the *sign* of a fermion's amplitude. In this way, we do not offend against the rotational symmetry of space; after all, we do not observe the amplitude of a process, but only its *probability*, which is the *square* of the amplitude. Since $(1)^2 = (-1)^2 = 1$, a change of sign of the amplitude makes no difference to the probability.

Now what happens when we superpose two fermions? There are two possibilities: either the fermions can be distinguished, or they cannot. If they are distinguishable (for example because their states have different quantum numbers, momenta or positions), nothing remarkable happens, because we have argued all along that there is no interference between distinguishable alternatives (remember the two-hole electron interference experiment).

On the other hand, if the fermions are *not* distinguishable, we obtain interference between the two alternatives: one direct process and an exchange process, which is the same as before but with the fermions interchanged (Fig. 8.21). It must be emphasized that indistinguishable is meant *very* literally in the quantum world: small things can be *identical*, big things cannot. Identical twins aren't, unless they are electrons' kin.

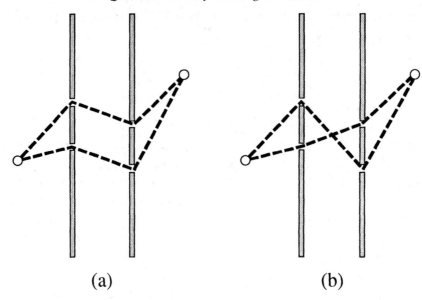

        (a)   (b)

Fig. 8.21 Example of a direct process and its exchange. At (a), we see two Feynman paths connecting two points. At (b), the intermediate legs of the journeys have been exchanged. Notice that the pieces of the paths outside the region between the screens remain the same, but the phases along the paths are different.

If we call the amplitude of the direct process $a \uparrow$ and that of the exchange process $a \downarrow$, then the quantum rules demand that they be linearly superposed to obtain the total amplitude of the process.† But we have seen that a single fermion can be exchanged for a negative-amplitude copy of itself without changing its probability distribution. Accordingly, we must add the direct and exchange processes by the prescription (total amplitude) = (direct amplitude) – (exchange amplitude) or, in symbols, $a = a \uparrow -a \downarrow$. This requirement is called the `law of spin and statistics`; spin comes in because all fermions belong to the same spin class (we will see in a moment what the law of spin and statistics says about bosons). If we apply this law to two indistinguishable fermions, we see that their superposition gives amplitude zero! Thus, we conclude that *identical fermions cannot exist*.

This statement of absolute intolerance is the `Pauli exclusion principle`: if we have a fermion in a given quantum state, then that state is thereby fully occupied. It is impossible to place a second fermion in it with the same state vector. This property forces fermions to aggregate

† A beautiful and detailed account of the connection between spin and quantum amplitudes can be found in Feynman's essay *The Reason for Antiparticles*, the first of two articles in *Elementary Particles and the Laws of Physics*, by R.P. Feynman and S. Weinberg.

into finite-sized lumps when piled up together, and it prevents them from building up large-scale coherent fields when exchanged as virtual particles. This behaviour is responsible for our *identifying fermions as matter* in the everyday world.

Contrariwise, it can be shown by an argument along the above lines that bosons have no objection at all to being in the same quantum state. The superposition rule of direct and exchange processes for indistinguishable bosons is $a = a\uparrow + a\downarrow$, which is the law of spin and statistics for bosons. Accordingly, the probability of having two identical bosons is *larger* than that of having these bosons separately. The probability $P = a^2 = a\uparrow a\uparrow + 2a\uparrow a\downarrow + a\downarrow a\downarrow$ is greater than $a\uparrow a\uparrow + a\downarrow a\downarrow$ due to the interference term $2a\uparrow a\downarrow$. It is this property that induces bosons to aggregate together in the lowest allowed energy state, without building up finite-sized lumps. On the other hand, bosons do build up large scale coherent fields when exchanged as virtual particles. This is the behaviour that is responsible for our *identifying bosons as forces* in the everyday world.

A space-time point can be occupied by more than one particle due to the possibility of superposition. Not only can more particles occupy one state, but there is a whole class of particles where such behaviour is in fact preferred (Bose–Einstein particles or bosons). On the other hand, there is another class where this is strictly excluded (Fermi–Dirac particles or fermions).

In summary we have:

property	*fermions*	*bosons*
amplitude addition	$a\uparrow - a\downarrow$	$a\uparrow + a\downarrow$
spin	$n\hbar/2,\ n = 1, 3, 5, \cdots$	$m\hbar,\ m = 0, 1, 2, \cdots$
statistics	intolerant	gregarious
examples	electron, neutrino, muon,	photon, W^+, W^-, Z,
	proton, neutron, quark	gluons, pions

The law of spin and statistics emphasizes the vital importance of spin for the behaviour of particles. Compare this with motions on a large scale: a giant object such as a star can rotate, but this has no influence at all on its motion through the Galaxy. Contrariwise, the spin of a puny electron is so important for its motion that it rules our very lives, as will be seen in the next chapter.

By the way, the entire argument leading to spin and its properties shows to what an astonishing degree the three-dimensionality of space is woven into the rules that govern our Universe. This might lead to the suspicion that

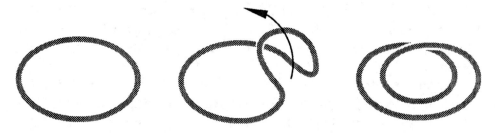

Fig. 8.22 Twisting the topology of a boson phase loop to obtain a doubled-over fermion loop. Note that the path can be made into a doubled loop only by moving it out of the plane of the loop (black arrow).

the existence of other aspects of quantization (for example, the difference between gravity and electromagnetism, or the quantization of electric charge) is related to the dimensionality of the vacuum. We will return to this briefly in Chapter 14, but as a preview, let us consider how exclusive the two spin classes are. Can a fermion ever become a boson, or vice versa? At first sight, this would seem to be utterly out of the question. The reason can be grasped by reconsidering the shape of the phase paths around the equator of spinning particles.

To change a boson into a fermion, we must change the *topology* of its phase path (Fig. 8.22). You can see how this works by placing a loop of string on a table, and trying to deform it from a circular (boson) shape into a double loop (fermion) configuration. You will notice immediately that this is impossible unless you lift half the loop off the table, twist it in space, and put it back on the other half. Thus, we find that unless we can step out of space into another dimension, we can never change a boson into a fermion, or the other way around. In fact, it can be shown formally that, unless our Universe has extra freedom beyond space-time, spin can only change in whole units of $\hbar$.

8.10 Polarization

Spin behaves like a classical rotation in some respects. As we have seen, the unit of spin is an amount of angular momentum, which is surely a rotation-related quantity. Also, the spin can have a certain orientation in space, called `polarization`. But there are important differences from the picture of rotation in classical mechanics. Because spin is a quantum degree of freedom, it behaves according to the quantum principle when observed: if you measure the spin of a particle in a fixed, but otherwise arbitrary, direction (let us call it the z-direction), either you observe a whole multiple of $\hbar/2$ or you observe nothing at all. Thus, with respect to any given direction, a particle can assume only a fixed number of possible polarizations.

Let us express the spin angular momentum of a particle as $S = s\hbar$. Then the spin of a boson is an integer: $s = 0, 1, 2, \cdots$, while the spin of a fermion is half an odd integer: $s = \frac{1}{2}, \frac{3}{2}, \frac{5}{2}, \cdots$. A particle with spin zero is called a `scalar particle`, because the absence of spin implies that one number is enough to describe the orientation of the particle. A particle with spin $\frac{1}{2}$, together with its antiparticle, forms a state vector with four components called a `spinor`. The name `vector particle` or `vector boson` is used expressly for a particle with spin 1. Bosons with spin 2 are called `tensor particles`, and bosons with higher spins are simply indicated as spin-3, spin-4, and so forth. A similar indication is used for fermions with spin $\frac{3}{2}$ and above.

The allowed orientations of the spin vector are restricted by the quantum rules, as follows. If a particle has spin s, the spin projected in the z-direction (called s_z) can go from s, via $s-1$, $s-2$, $\cdots$, to $-s$. The fact that the observed spin must change in jumps of 1 (and not $\frac{1}{2}$) follows from the fact that a particle cannot change from being a boson to being a fermion, or conversely.

If the observed spin projection is exactly in the z-direction, we have two possible cases: $s_z = s$ or $s_z = -s$. If the chosen direction is the direction of motion, these special polarizations are labelled `right-handed` (R) and `left-handed` (L). This designation is a leftover from the analogy with a classical spinning body: if we take a ball with right-handed spin in our right hand so that the fingers lie along the equator pointing in the direction of the rotation, then the thumb points in the direction of motion. Using the same analogy, in the special case where a vector particle has $s_z = 0$ it is said to have `transverse spin`. Which way a particle spins with respect to its direction of motion is sometimes called `handedness` or `helicity`; thus, there is R-helicity, L-helicity, and helicity zero. Upon counting from s to $-s$ in steps of 1, we see that there is a grand total of $2s + 1$ of possible values of the spin projection s_z.

For future reference, it should be noted that it does not matter how we specify the special direction in space-time on to which we project the spin. Any fixed direction can be a reference for polarization, as long as it is well defined: the direction of motion, the line connecting Sirius and Arcturus, the position of your sunglasses, or what not. The choice of a special direction is called `fixing a gauge` (we will see later that the fixing of a gauge plays a crucial role in our understanding of quantum interactions). For actual calculations, however, some choices are more practical than others. One choice is to describe the polarization as (L- or R-) helicity. Another is to describe the polarization perpendicular to the direction of motion. The two degrees of freedom in the case of a spin-$\frac{1}{2}$ fermion, like an electron or a proton, are usually called up and down spin (but note that the direction in space which you want to call 'up' is totally arbitrary).

Because spin is the offspring of relativity, we expect that the interactions among particles with spin can be described in a way that is obviously Lorentz symmetric. This turns out to be the case, but the details are messy; therefore, I will restrict the discussion to simply accounting for the number of degrees of freedom that a particle has. We recall that the number of degrees of freedom of a Lorentz symmetric object is a power of 4, because space-time has four dimensions. Therefore, we can have $4^0 = 1$, $4^1 = 4$, $4^2 = 16$, and so forth. Thus, a Lorentz symmetric object is either one number, or a string of four numbers, or an array of 4×4 numbers, and so on. The aim of the game is this: can we devise an assembly of particles such that their number of degrees of freedom is a power of 4?

First, a particle with spin 0. All we need is one number, namely the probability amplitude of the particle. This scalar particle is readily identified with the first case mentioned above: one number is enough, and 1 is the zeroth power of 4: $4^0 = 1$.

Second, a particle with spin $\frac{1}{2}$. This gives two numbers: one amplitude for the right-handed polarization and one for the left-handed (or one for spin-up and one for spin-down, if you prefer it that way). But 2 is not a power of 4! However, things can be made to work if we add the corresponding antiparticle; then its two polarizations provide the required two extra numbers. The particle–antiparticle pair is identified with the case where we have a string of four numbers, i.e. the first power of 4: $4 = 4^1$. Notice how Lorentz symmetry forces us to treat particles and antiparticles as if they were one object. This is precisely what we should expect from the preceding discussion of time ordering, and historically this is the way in which Dirac predicted the existence of antimatter.

The occurrence of two quanta, a spin-$\frac{1}{2}$ particle and its antiparticle, in *one* Lorentz-symmetric object explains why such particles are always created in pairs. In the Feynman diagram vertex, we must then have *one* spin-1 boson coupled to *two* spin-$\frac{1}{2}$ fermions (Fig. 8.9, Fig. 8.10). We can never have one spin-$\frac{1}{2}$ particle coupled to two spin-1 particles at a vertex.

Third, a particle with spin-1. This gives four numbers for the spin: one amplitude for the right-handed polarization, one for the left-handed, and two for the two mutually perpendicular transverse polarizations. Since the choice of the z-direction mentioned above is arbitrary, the choice of the orientation of the transverse components does not really matter, but unless we include it we cannot proceed with the calculation of physically useful numbers. The prescription of the direction is called `fixing a gauge`, and the fact that it doesn't matter what direction we take is called `gauge invariance`. Subject to gauge invariance, the particle with spin 1 is described by a string of four numbers, which can be Lorentz symmetric because it is a power of 4.

8.11 No mass, but as fast as they get

Apparently, nothing stands in the way of our using particles with spin in the description of relativistic quantum forces. But there is a troop of potential troublemakers: particles with rest mass zero. That such particles can exist, provided that they move with the speed of light, follows from Lorentz symmetry. I will not reproduce Einstein's famous calculation here, but we can get an idea of its main results.

The Lorentz symmetry keeps the Minkowski distance $x^2 - c^2t^2$ in space-time constant. Now if we want to construct a theory of mechanics that obeys this restriction, we must expect that the basic mechanical quantities (namely energy E and momentum p) will mix with each other in a way that is analogous to the mixing of space and time in the Minkowski recipe for the space-time distance. And indeed they do: Einstein showed that if $x^2 - c^2t^2$ is invariant, then $E^2 - c^2p^2$ must also be constant, and in particular $E^2 - c^2p^2 = m^2c^4$. There are two interesting special cases. One, if the momentum of the particle is negligible compared with mc: in that case we have $E^2 = m^2c^4$, which implies $E = mc^2$ or $E = -mc^2$. The first possibility applies to particles, the second to antiparticles.†

The other special case occurs in the opposite extreme: if the momentum p is very much larger than mc, we obtain approximately $E = cp$. Thus, if we keep the momentum fixed and let the mass m go to zero, we still obtain a mechanically possible connection between energy and momentum. Now if we apply De Broglie's prescriptions $E = 2\pi\hbar f$ and $p = 2\pi\hbar/\lambda$, we see that zero mass particles obey the relationship $f = c/\lambda$, or, alternatively, $c = f\lambda$. In words: (speed) = (frequency)×(wavelength). But this is precisely the requirement for a wave that propagates with the speed of light!

Therefore, particles with zero mass can exist, provided that they move with the speed of light. What can we say of the spin of such particles? To

† Dirac described antiparticles as objects with negative energy. If a particle is likened to a raindrop, then a Dirac antiparticle is like a bubble in the ocean. The water surface corresponds to the state of zero energy. In the Feynman picture that I have used throughout, an antiparticle must have a positive energy.

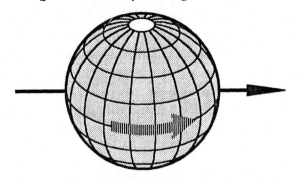

Fig. 8.23 Transversely spinning ball. The motion of the surface is indicated by the striped arrow; the spin axis is perpendicular to the velocity.

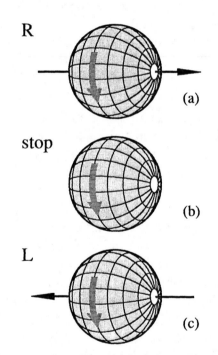

Fig. 8.24 Changing the polarization of a massive particle by overtaking it. (a) When we place the fingers of our right hand in the direction of the arrow, the thumb points in the direction of the velocity. This is called right-handed spin. (b) The particle stands still with respect to an observer moving with the same velocity. (c) When the observer moves faster than the particle, its apparent velocity is the reverse of that shown at (a). This particle is then seen to have left-handed spin, even though its sense of rotation has not changed.

make the discussion easier, let us fix a gauge in such a way that the axis along which we measure the spin (the arbitrary z-direction) is the same as the direction of motion (Fig. 8.23). This choice is called the `radiation gauge`. Now let us consider again the discussion at the beginning of this chapter. There, we shrank a particle to zero size while keeping its angular momentum constant.

Suppose that the particle is transversely polarized; if we look along its spin axis, one side of it would be moving in the direction of the particle's motion, whereas the other side would be moving in the opposite direction. But if the particle is massless, this cannot be: it must move with the speed of light in its entirety, so on both sides of the line of motion the speed must be c. Therefore, the difference of the motion of the two sides of the particle is zero, and so it cannot be rotating: its angular momentum in the transverse direction is zero. We conclude that a particle with zero rest mass can only have *two* degrees of spin freedom; in the radiation gauge, these are the left-handed and right-handed polarizations. Moreover, a zero mass particle cannot change its handedness, for we cannot slew it around without passing the forbidden transverse polarization region.

We can change the handedness of a massive particle (Fig. 8.24), simply by moving so fast that we overtake it: with respect to us, the particle will then move in the other direction, so that its handedness is reversed. But with massless particles that sort of trick is impossible, for massless particles must move with the speed of light and can never be overtaken: the speed of light is maximal.

9

Life, the Universe and everything

✩

9.1 Anatomy of an atom

In the second half of my story, the three strands of relativity, quantization and symmetry will be braided together. Before we embark on this journey, let us take a rest and consider some ways in which the properties of the small-scale quantum world govern our world at large. Most people wonder about the relevance of subatomic studies for our understanding of ourselves. Quite a few popular accounts of particle physics contain a heavy dose of reflections upon Larger Issues and related Deep Thoughts. I will try to avoid that, but it cannot be denied that the connections between the subatomic scale and our daily world are strong and direct, albeit invisible to our large-scale senses.

Let us first consider atoms, the building blocks that have the most direct significance in our daily lives (in the remaining chapters we will discuss the structure of the particles of which atoms are made). For simplicity and definiteness, and because we have encountered it in previous discussions, I will mainly speak of the hydrogen atom. Its nucleus is a proton, having one unit of positive electric charge. A neutral hydrogen atom is formed when an electron, bearing one negative unit of electric charge, binds to it. The binding makes the atom electrically neutral. We will see in Chapter 10 that electric charge neutrality makes the atom perfectly symmetric, in the sense that it is a singlet state of a certain symmetry group, just as a sphere is a singlet under rotations in space.

Because the mass of the proton is very large compared with that of the electron (1840 times as heavy), the inertia of the proton dominates the atom, and we can pretend that the proton is standing still. The electron cannot have an arbitrary distribution in space when bound to the proton. If the distribution is to be stationary, i.e. independent of time, the electron must be

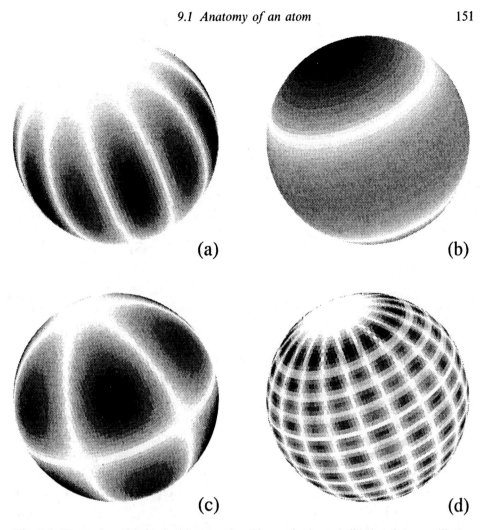

Fig. 9.1 Examples of spherical harmonics. Shown is the amplitude of the oscillation as a greyscale: white is low, black is high. The white bands are nodes, places where the spherical harmonics are zero. (a) $l = 6, m = 6$; (b) $l = 2, m = 0$; (c) $l = 4, m = 3$; (d) $l = 20, m = 10$. (Images courtesy of John Telting and Coen Schrijvers, Astronomical Institute, University of Amsterdam.)

in phase with itself everywhere, or else it would interfere destructively with itself. In other words, the wave pattern must be a standing wave.

This obliges the phase pattern of the electron to behave in a strictly ordered, quantized fashion: the zones are separated by a whole number of meridians and parallels which form the nodes of the phase pattern (Fig. 9.1). The meridian-and-parallel lines are not quite like the regular coordinate grid we see on an Earth globe, but are spaced in such a way that they cut patches with approximately equal surface out of the sphere.

Each piece of space that is bordered by nodal surfaces then encloses a roughly equal fraction of the electron's mass. In this way, the electron is evenly distributed on the sphere. The resulting partitionings of space are called `spherical harmonics`. These patterns produce all possible electron distributions such that each one can be rotated to coincide with itself. The spherical harmonics form a *representation of the rotation group*. They turn out to be precisely the allowed shapes of the electron distribution.

The nodal network on the surface of a sphere is built out of two families of lines: parallels and meridians. Accordingly, we have two quantum numbers associated with angular momentum: one for the parallels, called l, and one for the meridians, called m. The fact that there are two is a reflection of the fact that the uncertainty relations allow only two independent components of angular momentum. In the above, we saw that these could be taken as the total angular momentum J and the z-component J_z.

Each such pattern has a definite combination of energy and angular momentum associated with it. Roughly speaking, one can say that the higher the number of nodal lines, the larger the associated energy and angular momentum of that distribution. This follows from the uncertainty principle. If the wave amplitude is zero (namely at a wave node; remember the guitar string), then we know with certainty that the particle is *not* there.

Thus, the *more nodes* there are, the *less uncertain* we are about where the particle might turn up in space. Suppose that the number of nodes along a circumference on the sphere is l. More nodes on the surface of the sphere means that ϕ is less uncertain, because there are then many directions in which the particle is surely not to be found. In fact, we can estimate that $\Delta\phi = 1/l$. We remember that the angular momentum J and the angle ϕ are complementary, so that $\Delta J \Delta\phi > \hbar$. Thus, more nodes means a higher ΔJ, which must increase at the same rate that $\Delta\phi$ decreases. In this way, we find that the overall number of node lines on the sphere is proportional to J: if we cut the sphere along the z-axis, the number of nodes on this cut is proportional to J_z.

The possible energy states can also be calculated. We can make a rough estimate as follows. Suppose that the size of the region where the electron is found is r, and that the particle is distributed in a pattern with n nodes in the direction of r, i.e. away from the proton (the atomic nucleus). Because of the uncertainty relation, the momentum p of the electron must obey $p\,\Delta x = \hbar$. The argument is the same as in the case of the angular momentum: if the region of space where the electron is found has n nodes, then the amount of uncertainty of the electron's whereabouts is smaller by a factor n, because the electron is certainly not found at a node. Therefore, $\Delta x = r/n$, and $pr/n = \hbar$.

The electron is bound to the proton by Coulomb attraction, for which it is known that the binding energy E decreases with distance: $E \propto 1/r$ (we will see why later; the symbol $\propto$ means 'proportional to'). Because $E = \frac{1}{2}mv^2$ and $p = mv$, we have that $E = p^2/2m$, and so p^2 must be proportional to $1/r$ just as E is. We have already seen that the uncertainty relation requires that $pr/n = $ constant, which means that $p \propto n/r$. We insert this into the prescription for p^2, and conclude that the size r of the region where the electron can be found is proportional to the square of the number n of nodes of the electron distribution: $r \propto n^2$. If we put this value of r into the Coulomb attraction law $E \propto 1/r$, we finally obtain $E \propto 1/n^2$.

The energy of the electron in a hydrogen atom is inversely proportional to the square of the quantum number n. Where have we seen this before? Of course, in the Rydberg formula for the hydrogen spectral lines! The Rydberg formula can be recovered by looking at the energy transitions that can occur in the hydrogen atom.

The undisturbed atom would most likely be in the lowest energy state, called the `ground state`. Some idea of the appearance of the electron distributions can be obtained by showing the volumes occupied by the electron (Fig. 9.2).

Beware: these pictures do not show what an atom is or looks like, but they show the positions where an electron in an atom would be observed if we did a thousand experiments in which the location of the electron was determined. As should be clear from the preceding discussion, this is the only way in which we can visualize where an electron is. The cloud of dots indicates the density of 'electronness' in a given quantum state. In practice, the electron distribution drops off very rapidly beyond a certain radius, so that it is perfectly sensible to speak of an atom as an individual unit with a size r. Notice how very different this picture of electron wave patterns is from the impossible 'planetary model' of atoms!

If an appropriate amount of energy is added, the electron can be put into an `excited state` of higher energy (Fig. 9.3). Left to its own devices, the electron would return to the ground state by making spontaneous transitions, possibly via a number of intermediate states (Fig. 9.4).

Each transition from a higher to a lower state liberates a fixed amount of energy, namely the energy difference between the states, in the form of a photon. The frequency f of this photon is always the same for a given pair of states because $E = 2\pi\hbar f$: (energy difference between states) $= 2\pi \times$ (Planck's

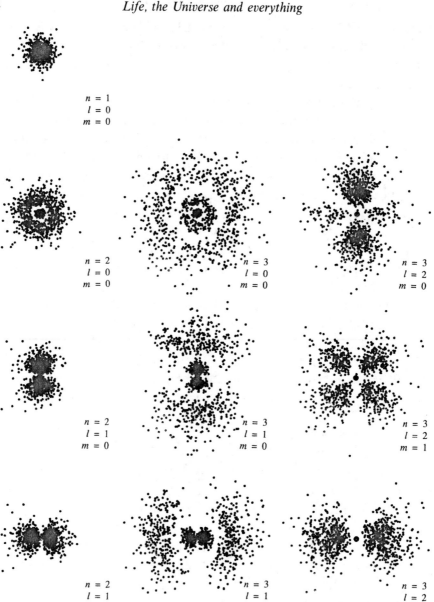

Fig. 9.2 Sections through electron distributions in hydrogen. Each dot marks the spot where an electron was found in a series of simulated experiments that determined the place of the electron. The density of the dots in any region then indicates the probability of finding the electron there. Top diagram: $n = 1, l = 0, m = 0$; leftmost column below that, from top to bottom: $n = 2, l = 0, m = 0$, $n = 2, l = 1, m = 0$, $n = 2, l = 1, m = 1$. The middle and right columns all have $n = 3$. The top diagram in the middle column has $l = 0, m = 0$. The two diagrams below that have $l = 1, m = 0$ and $l = 1, m = 1$. The rightmost column has $n = 3, l = 2$ and, top to bottom, $m = 0, 1, 2$. All diagrams are rotationally symmetric about their vertical axes.

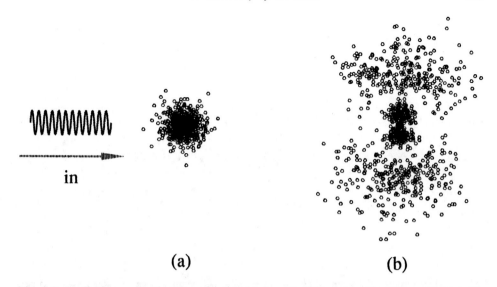

(a)   (b)

Fig. 9.3 Radiative excitation from the $(n, l, m) = (1, 0, 0)$ state of hydrogen to $(3, 1, 0)$. The incoming photon, symbolized by a wave coming from the left, is absorbed by the atom in (a). After the absorption process is over, the electron is distributed as shown in (b).

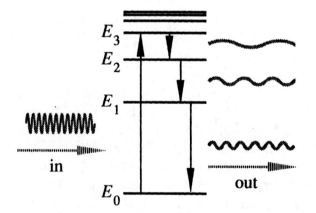

Fig. 9.4 Examples of radiative cascades in hydrogen from $(4, 0, 0)$ via $(3, 0, 0)$ and $(2, 0, 0)$ to the ground state $(1, 0, 0)$. The horizontal lines indicate the energy levels (compare with Fig. 2.7). The incoming photon is symbolized by a wave coming in from the left; the various outgoing photons have different wavelengths, because they correspond to transitions between different energy levels.

constant) × (frequency of photon). Thus, the hydrogen spectrum consists of a highly characteristic collection of possible spectral lines. From the chapter on spectra, we remember Ritz's law of the combination of spectral frequencies: if f_1 and f_2 are frequencies of spectral lines, then there are also lines at the frequency $f_1 - f_2$ and usually also at $f_1 + f_2$. This rule, which stumped classical mechanics, is now seen to be a consequence of the existence of internal energy states in an atom; to each Ritz term there corresponds one energy state. Thus quantization succeeds where classical mechanics failed.

Using the expression for the electron energies in hydrogen, we can explain the Rydberg formula $E = R(1/k^2 - 1/n^2)$. As originally discovered, this was a purely empirical relationship. But now we have seen that the internal energy levels are proportional to $1/n^2$; since it is the *difference* between two energy states that is emitted as a photon, we obtain the Rydberg relation directly as the difference between $1/k^2$ and $1/n^2$. The value of R, the Rydberg constant, was first calculated by Bohr, and corresponds exactly to the experimentally determined value.

9.2 Collapse prevention

The flavour of Bohr's argument is perhaps worth noting, because it shows uncertainty in action nicely. You may recall that, in the case of Newtonian gravity, the existence of stable objects proved that there are forces other than gravitation. The force of gravity is always attractive, so that any object that is composed of many particles which interact by gravity alone must inevitably collapse. It may take a long time, because a system containing angular momentum – such as the Solar System – can resist collapse until all the angular momentum has been transferred to other matter. But ultimately, the fact that all things always fall down bears in it the seeds of inevitable gravitational collapse, unless there are other forces that hold things up. How, then, can there be stable objects in the quantum world? How can a purely attractive force, such as that between an electron and a proton, be responsible for the stability of a bound state such as a hydrogen atom?

The answer is: quantum uncertainty, sometimes aided and abetted by Pauli exclusion. Suppose that the attraction between the hydrogen nucleus and its attendant electron squeezes the latter into a region of space with characteristic size Δx. Then the uncertainty principle demands that the momentum of the electron be indeterminate by an amount of the order of $\Delta p = \hbar/\Delta x$. Thus, the more we squeeze the electron in space, the more indeterminate the momentum is; the bigger the momentum uncertainty, the more likely it is that the electron can escape, no matter how strongly it is attracted to the proton.† Thus, the attempt to confine the electron to a small region of space makes it more likely that it will be able to escape!

Let us see how this works in the case of the hydrogen atom. The attractive Coulomb force between an electron and a proton, at a distance Δx apart, is (force) = (charge)^2/$(\Delta x)^2$ (this has the same form as Newton's law of gravity). Because force multiplied by distance equals energy, the energy associated with Coulomb attraction is (energy) = $e^2/\Delta x$, if e is the unit of electric charge. Now we know that (energy) = (momentum)^2/2m, where m is the mass of the electron. Because the uncertainty principle demands that the electron's momentum must be at least $\Delta p = \hbar/\Delta x$, we find that (energy) = $\hbar^2/2m(\Delta x)^2$. If we use the Coulomb value for the energy, we obtain $e^2/\Delta x = \hbar^2/2m(\Delta x)^2$. A bit of algebraic shuffling then gives $\Delta x = \hbar^2/2me^2$.

In other words, the smallest region in space to which the electron can be confined by the Coulomb attraction has a characteristic size Δx, given by the above expression (a careful calculation shows that $\Delta x = 2\hbar^2/me^2$, quite close to our rough estimate). *The uncertainty principle prevents the electron from collapsing on to the proton*! Notice again how different this is from orbiting electrons. Mercury doesn't crash into the Sun because of the planet's orbital angular momentum. But an electron in an S state has no angular momentum at all, yet it does not crash into the atomic nucleus. Any attempt to confine it to the very small piece of space occupied by the nucleus makes the momentum of the electron so uncertain that it can easily escape the nuclear electrostatic attraction.

As in all quantum calculations, this is only a statement of probability. What we have seen above is that it is *very unlikely* that an atomic electron collapses on to the nucleus. But occasionally such a collapse can be expected to occur, and in fact it does. Some nuclei, especially those that contain a relatively high number of protons, can grab one of their attendant electrons, and lower their nuclear charge by the inverse-beta process, in which a proton and an electron merge to form a neutron and an antineutrino: $p+e^- \rightarrow n+\bar{\nu}$. This is called K-capture. We will reconsider processes of this type, called weak interactions, below.

---

† Assuming that the force between the particles does not increase when the distance between them becomes larger. We do not yet know of any force that increases with distance, but in the next few sections we will discuss the fact that the colour force at large distances is independent of distance. If there should exist a force that increases faster than the first power of the distance between the particles, the present argument about the confinement of a particle would not apply.

9.3 Atomic symmetry

We conclude that the *symmetry is visible in the hydrogen atom*, albeit indirectly. The rotational symmetry of space imposes certain regularities on the electron distribution, because the electron must be in phase with itself: only those distributions are allowed that are representations of the rotation group on a sphere (namely the spherical harmonics). In the radial direction, the distribution is governed by the Coulomb attraction between the electron and the nucleus. These regularities are expressed in the energy levels, which show up in the frequencies of the spectral lines.

In the periodic table of the elements this regularity shows up as the famous 'period eight': chemical elements can be arranged in eight columns, in each of which elements with comparable chemical properties are grouped together. The number eight follows directly from the rotational symmetry in three-dimensional space. In other dimensions that number would be different. If we did not know that we live in three-space, we would be able to deduce that fact from the eightfold period in the table of the elements. Other such 'magical numbers' which occur in association with particles have in fact led to speculation that space-time has more than four dimensions, as I will discuss briefly in Chapter 14.

Actually, things are a little more subtle and instructive still: because of certain weak disturbances within the atom (e.g. the coupling between the electron spin and its orbital angular momentum, or the magnetic interaction between the proton and the electron), the energies of different angular momentum states do not exactly obey $E \propto 1/n^2$. Thus, *the rotational symmetry of the atom is broken*, and the electron distribution no longer behaves exactly as a representation of the rotation group. This makes the quantum numbers J and J_z observable, in the sense that each state with average energy $E \propto 1/n^2$ turns out, on closer inspection, to be split up into sublevels with energies that depend on J and J_z. The spectral lines originating from these sublevels form recognizable sequences in the atomic spectrum. The necessity to take J and J_z into account means that two more quantum numbers appear in the description of the hydrogen spectrum: the angular momentum quantum numbers l and m.

Even though these symmetry breaking effects are very subtle, they can have important consequences for us. For example, the energy difference between the hydrogen ground state with electron and proton spins aligned or with those spins oppositely polarized is exceedingly small: one trillion-trillionth of a joule, which is about equal to the energy liberated by a single iron atom dropped on to your table from a height of one metre (to produce five joules you would have to drop this book from about the same height). The radiation corresponding to this energy difference has a wavelength of 21 centimetres, and has been indispensible for radio-astronomical observations of hydrogen in the farthest depths of space. Thus symmetry breaking gives us a tool for the study of the Universe.

9.4 Quantization produces chemistry

If we consider a heavier atom, say oxygen, we find that its nucleus contains more protons (eight in this example). Thus, more electrons (eight, of course) are required to make the atom electrically neutral. One might jump to the conclusion that all of these would accumulate in the ground state, but that cannot be. Electrons are fermions, and the Pauli exclusion principle forbids the occurrence of identical fermions. Hence, the ground state can harbour only two electrons, provided they have different spin polarizations. Any others, for example the six more electrons needed to make the oxygen atom neutral, are obliged by the Pauli rule to occupy higher energy states, in which they are trapped because all lower energy states have been preempted by other electrons.

A higher energy state can harbour more electrons than the ground state, because higher states can have angular momentum. The ground state has no nodes, so it is spherical and has $J = 0$. But higher states have more nodes, and these can be distributed in space in several ways. Each one corresponds to a distinct spherical-harmonic pattern, and each pattern can hold two electrons (one with spin up, the other with spin down). Consequently, atoms of increasing mass are characterized by having more and more `electron` `shells` around them (see Fig. 9.2). The regularity of atomic electron distributions, which are imposed by the rotational symmetry of space, can be summarized by three quantum numbers: n, l and m. An electron distribution is fixed by its total pattern of nodes in space: in a radial direction as well as projected on to a sphere. In space, the nodes are surfaces that cut the region around the atomic nucleus as if it were a three-dimensional pie.

Because the total number of nodes is fixed for a given energy state, the quantum numbers cannot range freely. It can be shown that for any value of the `principal` `quantum` `number` n, the angular momentum number l must be less than n (but not negative). For example, if $n = 3$ then $l = 0, 1$ or 2. The absolute value of m must not exceed l; thus, if $l = 0$ one can only have $m = 0$, if $l = 1$ then $m = -1, 0, 1$, and if $l = 2$ one has $m = -2, -1, 0, 1, 2$. Counting the total, we find that there are n^2 different

The Periodic Table

1	2	3	4	5	6	7	8	9	10	11	12	13	14	15	16	17	18
H 1 (1)																	He 2 (4)
Li 3 (7)	Be 4 (9)											B 5 (11)	C 6 (12)	N 7 (14)	O 8 (16)	F 9 (19)	Ne 10 (20)
Na 11 (23)	Mg 12 (24)											Al 13 (27)	Si 14 (28)	P 15 (31)	S 16 (32)	Cl 17 (35)	Ar 18 (40)
K 19 (39)	Ca 20 (40)	Sc 21 (45)	Ti 22 (48)	V 23 (51)	Cr 24 (52)	Mn 25 (55)	Fe 26 (56)	Co 27 (59)	Ni 28 (59)	Cu 29 (64)	Zn 30 (65)	Ga 31 (70)	Ge 32 (73)	As 33 (75)	Se 34 (79)	Br 35 (80)	Kr 36 (84)
Rb 37 (85)	Sr 38 (88)	Y 39 (89)	Zr 40 (91)	Nb 41 (93)	Mo 42 (96)	Tc 43 (99)	Ru 44 (101)	Rh 45 (103)	Pd 46 (106)	Ag 47 (108)	Cd 48 (112)	In 49 (115)	Sn 50 (119)	Sb 51 (122)	Te 52 (128)	I 53 (127)	Xe 54 (131)
Cs 55 (133)	Ba 56 (137)	La 57 (139) •	Hf 72 (178)	Ta 73 (181)	W 74 (184)	Re 75 (186)	Os 76 (190)	Ir 77 (192)	Pt 78 (195)	Au 79 (197)	Hg 80	Tl 81 (204)	Pb 82 (207)	Bi 83 (209)	Po 84	At 85	Rn 86
Fr 87	Ra 88 (226)	Ac 89 (227) ••															

Lanthanide series*	Ce 58 (140)	Pr 59 (141)	Nd 60 (144)	Pm 61	Sm 62 (150)	Eu 63 (152)	Gd 64 (157)	Tb 65 (159)	Dy 66 (162)	Ho 67 (165)	Er 68 (167)	Tm 69 (169)	Yb 70 (173)	Lu 71 (175)
Actinide series**	Th 90 (232)	Pa 91	U 92 (238)	Np 93	Pu 94	Am 95	Cm 96	Bk 97	Cf 98	Es 99	Fm 100	Md 101	No 102	Lr 103

Fig. 9.5 Periodic table of the chemical elements.

states at each energy level E (which, as seen above, is proportional to $1/n^2$). Because an electron has spin $\frac{1}{2}$, it is a fermion with two degrees of freedom (spin up or down). Pauli exclusion then limits each atomic energy level to contain $2n^2$ electrons. Thus, atoms can be thought of as having electron shells that can contain $2, 8, 18, \cdots$ electrons in successively higher levels. The quantum numbers n, l, m which appear due to the properties of the rotation group in three-dimensional space thus produce 'magic numbers' of electrons in atoms.

At high values of the angular momentum quantum number l, there is an increasing energetic disadvantage in storing electrons there. The reason is that, at a given energy level, the principal quantum number n specifies the total number of spatial nodes of the electron distribution; thus, if l is small, the electron has few spherical-harmonic nodes and must therefore have many nodes in the radial direction. Accordingly, there is a rather high probability that the electron is found near the nucleus, so that low-l states are more strongly bound than those with high l (see the diagrams showing high-n electrons above). Therefore, the low-l states of levels with high n begin to be occupied *before* a previous shell is completed.

After the $n = 3$ states with $l = 0$, $m = 0$ and $l = 1$, $m = -1, 0, 1$ have been filled, the $n = 3$, $l = 2$ state is skipped. Instead, one finds that, first, the $n = 4$, $l = 0$ state acquires two electrons, and only thereafter does $n = 3$, $l = 2$ begin to fill.† The $l = 0, 1$ states can contain at most eight electrons altogether; this is the reason that the periodic table of the elements (Fig. 9.5) has *eight* principal columns. The $n = 3$, $l = 2$ state can contain ten electrons, so that the periodic table has a central block of ten columns. The number-eight rule has important consequences for chemistry.

The presence of the shells, mandated by Pauli exclusion, is the cause of the different chemical behaviours of the elements, and of the `periodic table` of these elements. If we add together all the wave patterns that belong to one energy state, but that have different angular momenta, the net result has no overall angular momentum (Fig. 9.6). Therefore, we end up with a perfectly spherical distribution. Its extreme symmetry makes this electron distribution extremely stable. The elements in which this occurs almost never participate in chemical reactions; with a rather feudal metaphor, they are called the noble gases (e.g. helium, neon, argon).

Bound combinations of atoms are called `molecules`. Those in which all angular momentum states are filled are very stable. This explains the `chemical affinities` that atoms have. For example, the element sodium (Na) has eleven electrons: two in the lowest energy shell, eight in the next higher one, and a single lone electron banished by Pauli to the outermost shell. Chemists say that sodium has 'valence one' because of this single electron. The element chlorine (Cl) contains 17 electrons: ten in the lowermost shells, just like sodium, but seven in the outer shell.

Thus, sodium and chlorine can join to obtain fully filled (and therefore fully symmetric) electron shells: sodium donates its valence electron to chlorine, so that it is left with two complete shells, and the Cl atom acquires three full shells. But then the Na atom is no longer electrically neutral: it misses one electron, and so has one positive unit of charge. Similarly, the chlorine has become negatively charged by one unit due to the extra electron. Thus, the atoms attract one another and form the neutral, stable, tightly bound molecule of Na Cl, which most people call salt. Next time you put some in your soup, think about how its properties are related to the rotational symmetry of space!

---

† This skipping makes it rather more difficult to recognize the quantum number sequence 2, 8, 18, 32,··· associated with the rotation group. The magic numbers do exist, but are not all readily apparent. If this also happens with the quantum numbers associated with other symmetries we will discuss, then it may be difficult to recognise the underlying groups in particle structure. For example, the masses of various particles may only *seem* to be irregularly spaced.

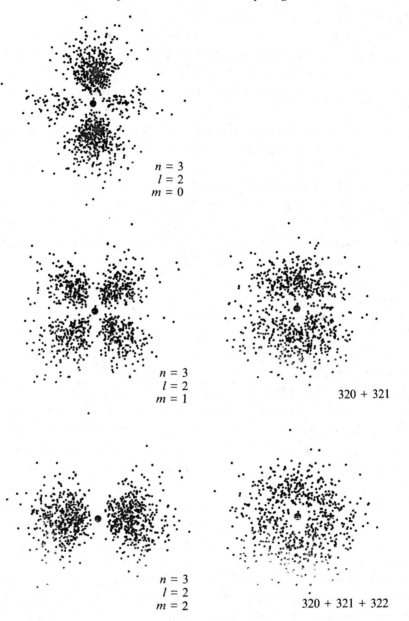

Fig. 9.6 The superposition of all possible electron states produces, for a given pair of quantum number values (n, l), a perfectly spherical electron distribution: its net angular momentum is zero. The leftmost column shows sections through the electron distributions with $n = 3$, $l = 2$ and, from top to bottom, $m = 0, 1, 2$. The pictures on the right are superpositions of these: top, $(3, 2, 0) + (3, 2, 1)$; bottom, superposition of all three.

9.5 The molecules of life

The sum of all electron distributions at a given principal quantum number
n (corresponding to a certain energy) add up to a distribution that has
no overall angular momentum. If J is known to be exactly zero, there is
no uncertainty at all in the angular momentum: $\Delta J = 0$. The uncertainty
relation for angular momentum, $\Delta J \, \Delta \phi > \hbar$, then implies that $\Delta \phi$ is infinite:
the particle distribution must be spherical because all information about
angular directions is lost. Thus, a completely filled electron shell is spherical,
the most symmetrical possible under rotations.

Most atoms do not have this configuration when they are electrically
neutral: they have unfilled shells. But to a certain extent they can enlist the
help of other atoms to complete their shells. We saw an example above. This
is the mechanism of `ionic binding`: the quantum advantage of making
spherically symmetric electron shells is so large that electrons are exchanged
between atoms, which thereby become electrically charged and attract each
other. Because ionic binding is due to the Coulomb attraction of opposite
electric charges, it is a long-range effect.

It requires energy to remove an electron from an atom because the attraction between the electron and the nucleus must be overcome. Thus, an
atom of carbon (C), which has two of its six electrons in $n = 1$ and four in
$n = 2$, generally finds it too difficult to scavenge four extra electrons from
elsewhere; nor does it readily let go of its own outer four. Instead, carbon
atoms are more prone to *sharing* their outer electrons with other atoms, in
such a way that two electrons (spin up and down) occupy the same spatial
pattern. A series of observations of the position of such a shared electron
would find it in (say) 35% of all cases in the $n = 2$ shell of the carbon atom,
and the remaining 65% in a shell of an adjacent atom (Fig. 9.7). Note that
this sharing is a pure quantum process: two particles share the same region
of space, and they are bound to two different nuclei at once (now try that
with billiard balls).

Because the electrons in this type of reaction, called `covalent binding`,
are paired in antiparallel polarizations, the atomic interaction has a *short
range* compared with the ionic bond. Thus, covalent binding acts only
when the atoms practically touch, i.e. their electron distributions begin to
overlap. Therefore, covalent binding between molecules is very sensitive to
the *orientation* of the participating molecules with respect to each other,
and is also very sensitive to the molecular *shape*. This condition makes the
covalent bond the basis of the chemistry of life.

The difference between the two binding mechanisms is not an either/or

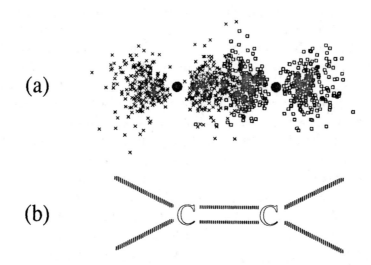

(a)

(b)

Fig. 9.7 (a) Overlapping electron clouds cause carbon atoms to bind together. (b) The usual chemical graphic notation for such a bond.

affair. An actual bond is never perfectly ionic or completely covalent, but falls somewhere in between. For example, hydrogen fluoride (HF) is 43% ionic, whereas hydrogen iodide (HI) is only 5% ionic. But it helps in understanding the chemistry of life to see the main distinction between these types of binding.

On the one hand there is the ionic bond, due to the attraction between opposite electric charges. The bond has a long range: it does not "snap" when the ions are pulled apart. The force between the ions gradually decreases as in Coulomb's Law. Accordingly, ionic molecules allow some flexibility in the arrangement of their constituent atoms. Due to the long range of the force the ionic bond depends a lot on the environment of the molecules, for example on the material in which they are dissolved.

The covalent bond, on the other hand, has a very short range. But once binding occurs this bond is by far the strongest. Separating or distorting such bonds requires a large amount of energy. Accordingly, molecules built with covalent bonds keep their constituent atoms at rigidly fixed positions in space. This is a consequence of the electron sharing in such bonds. As the images of Fig. 9.2 show, an atomic electron pattern has "handles" in only a few well-defined directions. This allows very little mismatch between the orientation of the electron distributions in adjacent atoms.

Because the ionic bond acts over such a long range it can cause long-range order in matter: ions make very good crystals (such as salt). But these extremely well-ordered structures that are rather boring from the point of view of living diversity: seen one salt crystal, seen them all. The covalent bonds typically do not produce such overly disciplined structures, but can create subtle diversity by being selective (short-range).

Now, consider the requirements of molecular architecture. How many configurations can a bag of marbles have? For practical purposes, infinitely many. The reason is that such a bag is physically large, even if it contains only a few dozen marbles. But a molecule, a bag of quantum marbles, behaves very differently. Because a molecule is small, the quantum rules allow comparatively *few arrangements* of its atoms, just as in the case of the fixed sequences of allowed electron distributions in the atoms themselves (and surely you remember that all electrons are *exactly* the same!) Here we get the first glimpse of the reason why living things occupy the middle ground between very small stuff (atoms) and very big things (planets).

Molecules have definite shapes, except for fleeting fluctuations. Large objects, such as a key, can appear to have a shape, but they can gradually wear down. Molecules cannot: if they change, they must change in discrete ways, from one definite shape to another definitely different one. You do not expect that a key fits the lock on your door one day, and that – due to wear and tear – it fits the lock on your neighbour's door the next! But a molecular key may suddenly snap to a state where it no longer fits its original molecular lock, but instead a totally different one. By the same token, molecules have structural strength because quantum laws do not allow them to change their shape into another that is too nearly the same. We literally live and die because of the fact that molecules have definite shapes (Fig. 9.8); for example, our immune system uses these shapes to recognize all manner of intruders and kill them as soon as possible.

Thus, molecules do not wear down: they are stable in time, unless either a specific molecular reaction intervenes, or a definite physical assault (e.g. a collision with a high-energy quantum) demolishes them, or quantum uncertainty flips their internal arrangement into something else. But the latter is rare: chemical energies are of the order of an `electron volt`, i.e. about 10^{-19} joule, which is the energy liberated when you drop a virus from a height of one millimetre. The uncertainty relation $\Delta E \, \Delta t > \hbar$ indicates that a spontaneous transition at that energy would occur only once every 30 million years.

For all these reasons, covalently bound molecules are recognizable and reproducible building blocks. Recognizable because of the close contact required for covalent bonding; reproducible because of the limited number

(a)

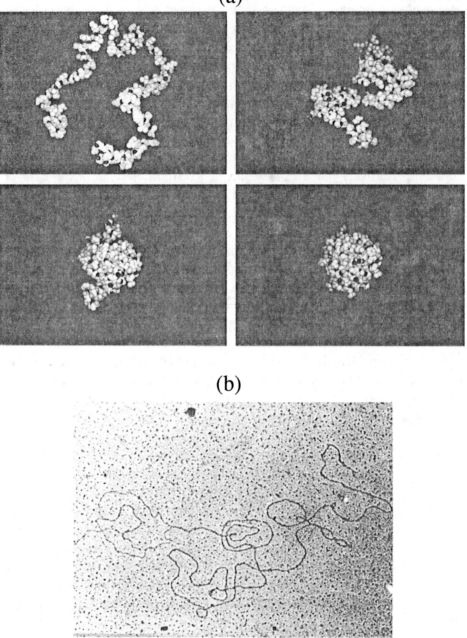

(b)

Fig. 9.8 (a) Small things have definite shapes: the strands of a protein fit together closely as the protein chain folds into its equilibrium conformation. (Computer simulation reproduced courtesy of Prof. F.M. Richards, Yale University.) (b) Big things do *not* have a definite shape, they can be deformed: the DNA molecule from the μ-virus lies spread out on an electron microscope preparation. (Photograph courtesy of Dr Nora Goosen, Gorlaeus Laboratory, Leiden University.)

of shapes a small thing can have. An insulin molecule in your blood can be *identical* to one of mine, whereas your bicycle cannot be identical to mine, even if it is the same make and model.

In *The Rhinoceros*, Ionesco's protagonists discuss at length whether a three-legged cat is still a cat. It is, because a cat is big. A three-legged cat will still be unmistakeably cattish, will purr, or whatever – it will function like a cat. But an insulin molecule with a single carbon atom missing is a totally different thing from a whole insulin molecule, even though the percentage missing mass is much smaller than the mass lost by the unfortunate cat. In fact, the insulin will most likely not work at all, or will indeed be poisonous. You can remove parts from the hapless cat right down to its last DNA molecule, and still, in a biological sense, it will be a cat. It will need major reconstruction, but the blueprint for that is the DNA. But as soon as you start to mess with that molecule, your modifications will make a real difference, and it will become increasingly un-catlike.

Covalent binding is an excellent way of sticking these building blocks together because the interaction has short range and the binding energy is high, which ensures that the chemical reactions can be highly specific.† The latter, of course, is required in a living organism, or else it would quickly degenerate to molecular nonsense.

Notice, by the way, that the above arguments are entirely general, since they depend only on the universal quantum behaviour. We should expect that life, wherever we find it, is based on structures that hold the middle ground of size: not as small as atoms, because these are so small that they never vary (seen one hydrogen atom, seen 'em all), and not as big as marbles, because these are so large that they have no definite shape (yes, I know that you think they are all spherical, but now put them under a microscope!) Large molecules occupy the middle ground between these extremes. The maximum size of an organism built from these blocks depends very much on circumstances; theoretically, there is no reason why a whale couldn't be as large as Mount Everest if the oceans were deep enough, but on this scale economics begins to play a role in selecting against organisms that are excessively expensive to build, maintain and control.‡

Some properties of life on Earth, for example the fact that it is based on the carbon atom, may or may not be general. Carbon, having four electrons

† See *The Molecular Basis of Life* by Haynes and Hanawalt, and, especially, the wonderful *Le Hasard et la Nécessité* by Monod. Also, one of the founders of quantum mechanics practically predicted the results of molecular biology about 20 years before they were discovered: see *What Is Life?* by Schrödinger.

‡ The way in which molecular mechanisms evolve is very well described in *The Blind Watchmaker* by Dawkins, and in several books by Gould, beginning with *Ever Since Darwin*.

in the $n = 2$ level, is the smallest atom that falls exactly between the two extremes in the Periodic Table, in the sense that it can either grab or give up four electrons to symmetrize its shell structure. Thus, carbon is an ideal atom for covalent binding, and in particular carbon can form highly symmetric covalent bonds with itself. In this way, it can form long chains of carbon atoms, with, every now and then, another one in between (like nitrogen or oxygen) to provide variety and to give molecular handles to which side chains can attach. For these reasons my money is on carbon as the universal basis of life. Another reason is that it is formed in abundance during the evolution of stars, whereas other valence-four atoms are much rarer (silicon ten times less abundant than carbon; all heavier valence-four atoms together are a thousand times rarer than carbon).

By the way, because an electron (spin-$\frac{1}{2}$) has *two* possible polarizations, the number of electrons in a filled shell is *even*. Hence there must exist atoms, such as carbon, which are exactly symmetric in the giving and taking of electrons, and thus are able to form long chains – a most amusing consequence of the rotational symmetry that divides particles into bosons and fermions!

It is extremely important to have an atomic species that can form chains, because only in this way can molecules attain the great diversity needed for adaptation to the large-scale world, and yet keep the essential smallness necessary for the preservation of well-defined quantum structure. When I read about molecular biology as a high school student, I was puzzled by the fact that structural molecules, such as proteins, are built up as a long chain that is then folded into a useful shape. Why not build it like a house or a ship, from all sides at once? Because biological molecules, to have enough diversity, must be biggish, as quantum systems go. Thus, their construction must be a compromise between quantum-small and classical-big. A molecule can venture into the realm of the large by extending in one spatial direction, as long as it stays slim in the others.

In between the classical realm of the big and the quantum realm of the small, there is a twilight zone which is the basis of life. When does the classical regime begin? That depends on how small an uncertainty you require. A statement such as 'with perfect certainty' then means 'classically'. But note that this is an imposed *a priori* condition, which cannot even be met, insofar as the whole Universe is a quantum mechanical object, and *absolute* certainty never occurs: the probability of something can be $1 - \epsilon$, but ϵ is never exactly mathematically zero. Thus, the question is meaningless unless you specify the amount of uncertainty below which you wish to call something 'classical'.

Life is based on the existence of this twilight zone in size. If it did not exist, if there were no absolute scale of large and small, then the code for new generations would be lost by erosion, as surely as a mountain will eventually be washed to the sea.

It is worth noticing that a randomly assembled molecule, such as might arise when a batch of atoms is bound any which way, is extremely unlikely to have the linear form required for the transfer of encoded information. More likely, such a molecule would resemble a mini-glob of atoms. The fact that linear molecules came to dominate is significant. This linearity is itself evidence that heredity and Darwinian selection must have played a decisive role in life from its most primitive beginning in the 'primeval soup'.

I am convinced that Mendelian inheritance follows from the quantum laws, too. You may recall that one of the surprising findings of the study of genetics (which probably caused Darwin many sleepless nights) is that particular traits in offspring are not a smooth blend of the traits of their parents. Rather, genetic inheritance is an either-or case: my father is a man, my mother is a woman, but I am a man and not 50% man and 50% woman. Similarly, my sister is 100% a woman. Biologists called inheritance 'particulate', a very apt term indeed. The quantum cause for Mendelian inheritance is this: a molecule is *not* a particle which is an average of its atomic ingredients. The object called H_2O is a water molecule, which is not in the least a blend made of two parts hydrogenlike and one part oxygenlike properties.

There is no such thing as an average in quantum mechanics (except in the technical sense of a chimaera superposition). A state is either entirely one thing, or wholly another. Any seemingly smooth distribution is really a pile of clicks, just as a sandcastle is a shape built of a trillion grains. Because our genetics is encoded on an atomic level, it is an either-or case, resulting in the remarkable 'particulate' properties of genetics.

9.6 Quantization and genetic coding

In spite of the somewhat hyperactive prose that the quantum laws often elicit from philosophers, the connection between quantization and life-and-all-that is quite genuine. In fact, Schrödinger wrote a prophetic book about the small-scale properties of life long before the discoveries of molecular genetics. We have just seen why life must be based on structures that occupy the middle ground in size, and that these are built with covalent bonds. But I suspect that the quantum laws have even more far-reaching consequences, namely in the mechanisms by which living things produce amazingly accurate copies of themselves. Molecules are recognizable and reproducible building blocks, and the intimate contact that covalent reactions require makes it easy to accept that there can be molecular recognition processes, in which molecules fit together like a hand in a glove. But can we place quantum constraints on the way in which they reproduce?

One can almost define life by its accretion, incorporation and organization of stuff from outside: growth and self-replication. Living objects can conceivably make exact copies of themselves by some contact-printing process; indeed, the reproduction of some basic molecules in living things proceeds in this way, and it is plausible that life started like that. But this replication mechanism is limited to structures of fairly small size because the reproduction of big things cannot be exact.†

There may be a selective advantage in being big, for example if one needs a sophisticated mechanism for self-repair or the gathering of materials to cope with the destructive influences of the outside world. Organisms can only combine big size and exact replication by replicating when they are small and then growing. But future generations cannot be folded up in the present generations that are their parents. Quantization reigns in the realm of the small, and below a certain size one cannot reproduce a miniature version of something much larger. A one-to-a-billion scale model of an aeroplane is physically impossible. In classical mechanics, there is no reason why you could not have been present, in miniature, in your mother's ovum, and your children as yet smaller copies within you. But the absolute scale of smallness imposed by the quantum laws implies that future generations are not enclosed in us as minipeople. And yet the extreme similarity between one generation and the next indicates that living things are, in fact, present in their ancestors. How?

If offspring cannot be present in their parents in actual form, then all that remains is that they are enclosed as a list of instructions, as a blueprint; in other words, in *symbolic* form. On a molecular level, the coding necessary for a symbolic image is possible because the quantum laws allow only a few very specific arrangements for the atoms in a molecule. Therefore, molecules are good data carriers: they are good bits of information, provided that they are neither as small and identical as atoms, nor as large and irreproducible as marbles.

By the way, this implies that, all over the Universe, the adaptation of larger organisms to their environment must go via *indirect* selection: because

---

† The superb fidelity of the biological replication process is well illustrated by Dawkins in *The Blind Watchmaker*.

offspring are only present in symbolic form in their parents, the environment does not affect them directly. Of course, gross damage can be done to genetic molecules, for example by chemicals or radioactivity, but that is beside the point here. The connection between a code and the thing it stands for is accidental: a string of letters that forms a perfectly polite English word may be libelous in Dutch. Thus environmental effects can act upon the physical form of an organism, and hamper or destroy those that are unfit, but they cannot read the genetic code and, say, delete all four-letter words that begin with an F.

What sort of shape could the blueprint molecule have? On the one hand, it must be very, very big in order to contain the enormous amount of information that is needed for the description of a large being. On the other hand, it must be very small, for all the quantum reasons elaborated above. These two conflicting requirements can only be fulfilled together if the coding molecule is much bigger in one direction than in another. Therefore, it cannot resemble a ball, but must be more like a sheet or a string. The safest form would be one in which it is extended mostly in one direction, and is small in the other two. Thus we arrive at the conclusion that the quantum laws require that procreation of larger organisms occurs via an intermediate, symbolic step in which the next generation is encoded in a molecule that is like a very long string or a ribbon (Fig. 9.9).

The large size of such a molecule is not without hazards, even though it is extended in one direction only. It may break into pieces, or be tied into a knot and re-splice itself in random ways. A large molecular string would do well to be rolled up carefully, in order to avoid disastrous tangles that could wipe out the meaning of its coded message. In fact, it is to be expected that any coding string will become subject to mutations, because of its size alone, as soon as its linear dimension becomes big in the quantum sense. On the other hand, these size-induced mutations could provide the initial impetus for the evolution of organisms bigger than the molecules of which they are made.

Of course you know that such molecules exist: the deoxyribonucleic acid, abbreviated DNA, that encodes you and me, and the ribonucleic acid or RNA that encodes some viruses (quantum theorists are not the only ones to use impossible words). The exact form and behaviour of these molecules is probably specific for life on Earth, but I think that the laws of quantum mechanics indicate that their *form* (long and thin) and their *function* (specifying future generations in symbolic code) are as universal as the quantum laws themselves.

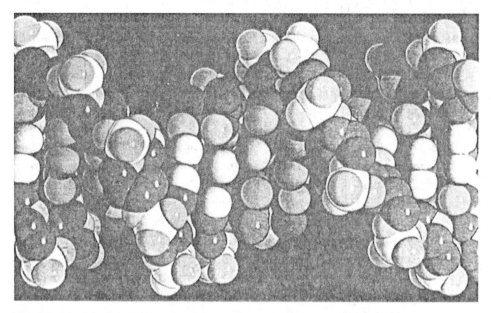

Fig. 9.9 Model of the molecule of deoxyribonucleic acid, or DNA, a genetic blueprint of life.

9.7 Will the Universe live forever?

If our Universe, at some time in the future, should reverse its present expansion and collapse, it could not harbour life forever: everything would be snuffed out in a Big Crunch. But astronomical observations indicate that the Universe will probably never stop expanding. Therefore, it will cool down indefinitely, but does that preclude life in the extreme future? Maybe not. The arguments above have been given for atoms that are made of electrons, protons and neutrons, but that is mostly by way of example. The entire chain of reasoning depends exclusively on general quantum principles, such as the uncertainty relation and on the law of spin and statistics, so it seems to me that life is possible as long as there are bosons and fermions.

The decrease of the temperature of the Universe is not necessarily an argument against future life because it is perfectly possible that there exists a superweak force that becomes apparent only at very low energy density. Our chemistry is based on the Coulomb attraction and on the pairing of electron spins, but it is by no means excluded that in the extreme future the force of electromagnetism would freeze out and leave a residual force of a different kind. Indeed, current speculation holds that this is precisely the way in which electromagnetism itself became manifest in the early Universe;

likewise, 10^{40} years from now it will seem that the few electron volts that power us and our microchips are gargantuan energies.

The danger lies not in the need for absolute amounts of energy, but in the availability of energy *differences*. If these did not exist, everything would be in equilibrium and life would be impossible. However, the Universe is expanding, and, as long as it does, it is not in equilibrium. It is totally unclear how that affects the formation of material structures, whether dead or alive, but in an endless Universe the possibilities are endless, too.

10

The physics of a tablecloth

✦

10.1 Picking up the trail

After this brief excursion into the world of the large, let us descend again into the quantum realm. In the following chapters I will try to bring together the main elements of the story, namely relativity, quantization and symmetry. We have seen how the first two lead to a picture of the subatomic world in which quanta are constantly exchanged between vertices in space-time. These quanta can be divided into two groups, the fermions and the bosons, which obey the Law of Spin and Statistics. One of the consequences of this law is the identification of fermions with matter and bosons with forces. We also saw that relativity and the boson–fermion classification can be attributed to certain regularities in the behaviour of space and time, called symmetries: conservation of angular momentum is tied to the fact that space does not change when it is rotated; the laws of relativity and the law of spin and statistics are consequences of the fact that space-time does not change when it is subjected to a Lorentz rotation.

It remains for me to argue that such symmetries are the all-encompassing organizing principle in the Universe, in the sense that they dictate the way in which fermions and bosons are coupled at a vertex. Thus, symmetries provide the final link in the story, because they prescribe the relationship of matter and forces. The remarkable thing is that all the known forces of Nature have been shown to obey the same mechanism. Therefore, physicists have conceived the hope that a single explanation for all forces will soon be discovered. Accordingly, the word 'force' appears in the title of this book in the singular!

We are now at a pivotal point at which it is discussed how a symmetry can govern the coupling of fermions (matter) to bosons (force fields). We will see that symmetries are so powerful that, given a set of fermions, we

174

can calculate exactly how many bosons can interact with them, and almost exactly what the properties of these bosons are. The key to this explanation is a simple tabletop experiment.

10.2 Mind your multiplets

In high school, I was told that the Universe was made of seven ingredients: protons, neutrons, electrons, the forces of electromagnetism and gravity, and space and time as an invisible graph paper on which the motions of the first three were plotted. Later, more items were added: the weak and the strong nuclear forces, and a bewildering array of particles – pions, muons, photons and what not. Their motions were described in terms of momentum, kinetic energy and potential energy. I am not likely to forget the feeling of disgust produced by this oppressive multitude. Why did we have scores of particles? Why rejoice when yet another little nipper was discovered? Why four forces? Why so many kinds of energy?

Then, in university lectures, simplifications began to appear. Space and time merged into space-time under the command of the Lorentz symmetry, which welded energy and momentum together in the energy-momentum vector. Space-time took its place as one of the kinds of stuff of which our Universe is made. Forces were quantized and potential energy was absorbed in the particle picture. Gradually, the scene simplified to a stark ballet of fermions and bosons, linked in space-time to form an infinitely divided mesh, coupled together at vertices according to the rules of symmetries that showed a strong resemblance to the rotational and Lorentz symmetries of space and time.

I have tried to give an account of the amazing progress that physicists have made towards simplicity in the description of the world's construction. Much more work remains, but let us try to see how, at present, the simplifying concepts of relativity, quantization and symmetry act in concert.

When presented with such a bewildering array of ingredients, one's first reaction might be: what do all these have in common? The question is similar to the one that was raised by spectroscopists who were confronted with the spectrum of hydrogen. Do we see any regularities, trends, classifications? In modern parlance, what we seek are `multiplets`, that is, sets of particles that show a family resemblance. An object that is not changed when subjected to a symmetry operation is called a `singlet` under that particular symmetry (e.g. a circle is a singlet under rotation). A pair of objects that turn into each other is called a `doublet`, and so on, with triplet, quartet and beyond.

To illustrate this kind of classification guessing game, consider an analogous situation in biology.† Suppose that we have the following list of observed beings (in alphabetical order):

blue-green alga	chimpanzee	human
bacterium	fish	starfish
cat	geranium	virus

Can we find any 'Linnean multiplets', i.e. sets that have biological properties in common? Certainly, for example the doublet made up of (alga, geranium): they are both photosynthesizers, using sunlight to make food. If photosynthesizers were otherwise indistinguishable, then the above pair would be a doublet under some sort of *photosynthesis* symmetry. Similarly, the pair (chimpanzee, cat) is a doublet for the *furry* symmetry, and (cat, chimpanzee, human) is a triplet for *hairy*. Then (cat, human, starfish) is a triplet for *predators*, (cat, geranium) is a doublet for *domestic* (maybe the human should be included to make this a triplet) and (fish, alga, starfish) is a triplet for *aquatic*. We have (cat, chimpanzee, fish, human) as a quadruplet of *vertebrates*. In the realm of actual symmetry groups, (starfish) is a singlet under *rotations* over 72°, unless the virus happens to be one of the icosahedral types, in which case it ought to be included. All entries except the geranium form an octet under the group *reflection*, because each has a central plane of symmetry. Thus, we see that the lion may lie down with the lamb as members of a mammal multiplet, but they won't in a predator multiplet.

None of this seems at all fundamental or compelling, although the occurrence of the geranium in *two* multiplets could suggest that it might be worth our while to domesticate the blue-green algae. There are many classifications and interpretations, and yet biologists have found that organisms such as those listed above can be profitably grouped in larger and larger multiplets (Fig. 10.1). These turn out to contain information about the way in which the multiplet members evolved. The smaller the multiplet, the later it separated out as a batch of species.

This biological exercise may be profitable for people who try to impose order on a real zoo, but in the zoo of particles the aim is a bit different. When we deal with real lions and real goats, we are concerned with large objects, the quantum behaviour of which is not readily apparent. But a very *very* small goat and a similarly small lion could conceivably be members of the same multiplet because of the possibility of quantum superposition. As mentioned before, any state can in principle be a superposition of others,

† A similar example can be found in Dawkins's *The Blind Watchmaker* in his discussion of 'clades'.

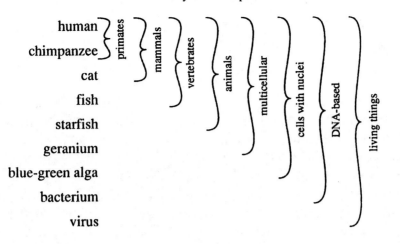

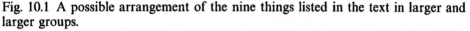

Fig. 10.1 A possible arrangement of the nine things listed in the text in larger and larger groups.

and so a lion and a goat can be a doublet under stirring in some abstract apprentice's cauldron. Thus, lion and goat can be specific instances of a more general beast; we saw that this mythical animal is described by the equation (chimaera) $= g \sin \theta + l \cos \theta$.

When we try to find multiplets of particles, there seem to be few surprises at first. The fact that the neutron is only 0.14% heavier than the proton suggests that we gather them as a doublet under some, as yet undefined, isospin group – the name is an echo of the spin up/down freedom of the electron, and we surmise that the proton is up and the neutron down in some sort of abstract space. The isospin group symmetry then turns a proton into a neutron, and conversely – rather like the chimaera.

But masses or electrical charges are not always good indicators for taking fermions together in a multiplet; for example, it will turn out that the electron and the neutrino can be assembled in a doublet under the 'weak SU(2)' group. There is as yet no unique or compelling way to classify fermions. In other words, the simplification process to which I alluded above is far from complete, and so is our understanding. However, we will soon see that the application of the symmetry principle has already produced striking results.

The process of invoking a symmetry group to derive the properties of a boson field coupled to a fermion multiplet proceeds as follows. First, we identify or postulate a set of N fermions that are observed, or expected, to act as a 'fundamental' multiplet. The number N is called the dimension of a symmetry group G, and the set of fermions is an N-plet under G. Second, we suppose that an operation taken from the group G, when applied to a member of the multiplet, turns this into some other member. Thus, the group elements serve to transmute one fermion into another. Every transmutation is interpreted as being due to the emission or absorption of a field boson. In our chimaera analogy, $N = 2$, and the fundamental doublet is (lion, goat). The symmetry operation is a rotation over the 'stirring angle' θ.

This sounds rather abstract, but we will see this procedure in action in several examples. To preview a particularly simple one: if we take the electron e^- to be a fermion that cannot be changed into anything else, then it is a singlet under some one-dimensional group U(1) (in the section on electromagnetism we will see that U(1) is the group of one-dimensional rotations). All that this group can do is turn the electron into itself; thus, there is only one boson associated with the group, called γ: the *photon*, which couples directly to the electron at a vertex. We have seen this kind of coupling† in all our preceding discussions of Feynman diagrams in electromagnetism.

Notice, by the way, that at this stage of the game we are only guessing at multiplets. Thus, the fermions we come up with cannot yet be considered elementary, primary, fundamental, or whatever. Remember that early spectroscopists classified the spectral lines of hydrogen as Sharp, Principal, Diffuse, Fundamental, or what not, and that these labels turned out to be physically meaningless. But we are doing something akin to what spectroscopists did, which culminated in the Rydberg formula: finding order and simplicity. Underlying the empirical Rydberg equation we discovered a basic rotational symmetry that governs the structure of the atoms. Likewise, we hope to discover symmetries that govern the structure and interactions of the particles of which atoms are made.

10.3 Twists and wrinkles

We saw previously that, by means of linear superposition, any quantum state can be seen as a composite of other states. Because of certain requirements of decency (for example that a state, if superposed with itself, cannot yield anything else but itself), the result of the superposition of two states (Fig. 10.2) can be written as (some state) $\times \cos \theta +$ (some other state) $\times \sin \theta$. Accordingly, if we want to devise a transformation that treats the individual

---

† See Feynman's superb book *QED*.

states as equivalent to each other under some symmetry, then that symmetry behaves like a rotation† with a `mixing angle` θ.

Using this superposition, starting with equal amounts (amplitudes) of lion and goat, we found that $\sin^2 \theta$ is the probability that a chimaera behaves like a goat, whereas $\cos^2 \theta$ is the probability that it behaves like a lion. Suppose, now, that the sorcerer's apprentice has to produce a real chimaera as a graduation project: that means that it must also contain a certain amount of serpent, in addition to its goatness and lionness. How is this superposition to be achieved? Well, a clever apprentice would notice that the undergraduate chimaera was made by rotating a pure goat state G and a pure lion state L in a cauldron, which suggests immediately that a pure serpent state S be added in such a way that the resulting superposition can be rotated smoothly in (G, L, S)-space. This would require stirring of the mixture in three rather than two dimensions, but that is merely a matter of cauldron design. The basic idea remains, namely, that the symmetry operation that turns G into L, L into S, or S into G, behaves like a rotation, albeit in a space of higher dimension.

Thus, if we pick a fundamental fermion multiplet with N members, we expect that the symmetry group that acts on this N-plet should behave as a rotation in some abstract N-dimensional space. In the (G, L)-space example, we had $N = 2$, and the rotation of a particle over the mixing angle θ literally makes the particle 'turn into' a different one! Some symmetries are actually connected with rotation in space (which creates the angular momentum of a particle) or rotation in space-time (Lorentz symmetry). Other symmetries behave like rotations, too, but not in ordinary space; rather, these symmetries are rotations about other directions than the axes of space and time. Apparently, the vacuum possesses more possible directions than those of space-time.

The similarity between the symmetries of Nature and simple rotations enables us to understand how a symmetry can produce a force field. This can be done in a tabletop experiment that is so simple and direct that I strongly recommend that you do this experiment right now.

Take before you, on a smooth table, a small tablecloth, or something similar (e.g. a large piece of aluminium foil). The material must be a uniform colour, without any patterns. Make sure it is quite smooth. We are not looking so closely that the individual fibers are visible, and we will pretend that the material extends to infinity: our tablecloth is a small piece

---

† Sticklers for detail will take exception to this: in fact, *any* group can serve as a basic symmetry. But I am attempting here to show what the hitherto most successful groups have in common, by exploiting their family resemblance to the familiar rotations.

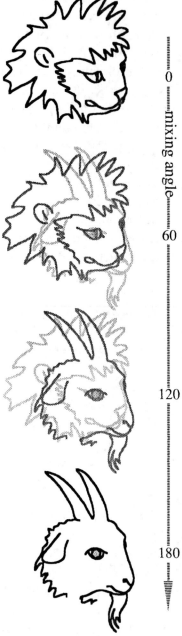

Fig. 10.2 Superpositions as a function of mixing angle. Top panel shows a pure quantum state (the lion), the bottom diagram shows another pure state (the goat). In between are two superpositions, with mixing angles 60° and 120°.

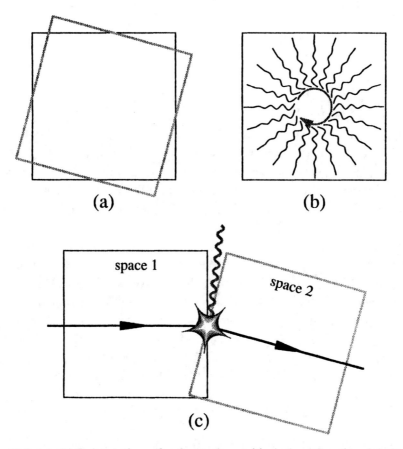

(a)   (b)

space 1

space 2

(c)

Fig. 10.3 (a) Global rotation of a featureless tablecloth. After the global rotation, the cloth has exactly the same appearance: it is globally symmetric under rotation. (b) A local rotation causes the cloth to wrinkle. The tablecloth is not symmetric under local rotations: there is a mismatch between the local twist and the cloth at large distances. This mismatch shows up as a wrinkling. These wrinkles correspond to the appearance of a field ('field lines') in the case of a local quantum symmetry applied to the vacuum. In a Feynman diagram, the wrinkle appears as a photon (c).

of an unbounded model universe. Now rotate the whole cloth through an arbitrary angle (Fig. 10.3). You will, of course, notice that any piece of the surface, when inspected individually, appears the same as before: the cloth is *invariant under global rotations*.

This part of the experiment seems thoroughly trivial, until you realize that it would be impossible, even in principle, to do something like this with the real Universe. Imagine that we want to perform a global symmetry transformation. Then we would have to let the symmetry act in all of space at exactly the same time. But this is impossible to do in reality because no signal can propagate faster than the speed of light. As we saw when discussing action-at-a-distance, the fact that the speed of light is finite and maximal means that any signal takes time to cross space. This led to the concept of a field and the idea that an interaction is the exchange of a field quantum.

Relativity tells us that we must abandon the idea that a symmetry can be global: because of the fact that nothing can travel faster than light, we cannot apply a symmetry in the whole Universe at once. It would appear, then, that the way out is to *accept only local symmetry rotations*, that is, a symmetry where the amount of rotation *differs* from event to event in space-time.

Return to the tablecloth before you. Put your finger on a point near the centre and give the cloth an arbitrary twist, keeping the edges of the cloth in place (Fig. 10.3). When you remove your finger, you notice that the piece of the surface you have just rotated still appears the same; at that one point, the cloth is invariant under local rotations. But in the vicinity of the twisted point, something has happened: *a spray of wrinkles radiates outwards from it*. The local twist cannot be connected smoothly with the undisturbed cloth at large distances: the difference must be patched up. Because of relativity, all symmetries must be local, and *any local symmetry creates a field*. The wrinkles are related to the 'field lines', as they were called in pre-quantum days.

This analogy may help you to understand the old problem: if an atom drops to a lower energy state and emits a photon, where was the photon before that? The answer, shown in the lower half of Fig. 10.3, is: the photon was in another world, another 'abstract space', and has become apparent at the juncture between the space containing the single electron. The U(1) twist connects the two spaces.

Again: a *local* symmetry is one which differs from point to point in space-time, or from point to point in some abstract general or internal space. A local symmetry twist creates a mismatch between the local area and the rest of the Universe. The mismatch is patched up by the appearance of a field.

10.4 Gauge twists and velpons

Suddenly, rotating pieces of universe doesn't seem so trivial any more! We have shown how a symmetry can be responsible for generating a field. Now in the quantum picture a field is built up from field quanta; and the exchange of a quantum produces a force. Thus, *any local symmetry creates a force*.

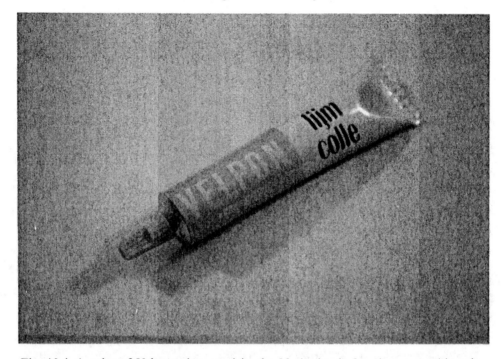

Fig. 10.4 A tube of Velpon glue, used in the Netherlands for almost anything; its effectiveness in capturing quanta remains to be proved.

The symmetries discovered so far behave in a special way, which is called a 'group' by mathematicians. The rotation-like symmetries introduced above are called gauge groups and the wrinkles that occur in the vacuum due to a local gauge transformation are called a gauge field. The quanta of such a field are gauge quanta, or, more precisely, gauge bosons. The force that corresponds to the exchange of gauge bosons can be considered as a binding agent, a kind of glue between the particles to which the bosons are coupled.

Each force has its own set of glue quanta. A generic name (other than the insipid 'gauge boson') for these does not exist in the professional literature. Therefore, I will succumb precisely once in this book to the temptation to name something, and use the generic name velpon, after one of the most common brands of glue in my home country (Fig. 10.4).

All known forces are due to gauge symmetries; in what follows, we will discuss which force goes with which symmetry. For the first time ever, we are beginning to see a profound similarity between the forces of Nature. The fact that these forces can be traced to a similar mechanism (local gauge invariance) does not necessarily mean that they are the same; but most physicists hope, and perhaps expect, that all forces of Nature are, in a deep sense, different aspects of the same phenomenon. As the Dutch physicist 't Hooft wrote: 'Thus if physicists have yet to find a single key that fits all the known locks, at least all the needed keys can be cut from the same blank.'

Some confusion might arise here because of different meanings of the word force. In the sense used in this book, the action of a force corresponds to a change of momentum due to interaction with the fields of electromagnetism, weak nuclear force, 'colour' or gravity. In the more general sense used in mechanics, force can be due to other quantum effects as well. For example, if we try to persuade a third electron to take up residence in the $n = 1$ state of helium, we will not succeed because of Pauli exclusion. Thus the third electron is repelled by a process that, seen on a large scale, acts as a repulsive force. Indeed, there are stars, called `white dwarfs`, that can contain 1.4 solar masses within a volume as small as Earth; in these, the pressure needed to counterbalance the pull of gravity is provided by Pauli exclusion among electrons. Similarly, the supercompact `neutron stars` have up to two solar masses in a sphere with a mere 10 kilometre radius, held up by Pauli exclusion among neutrons. Of course, one might argue that the law of spin and statistics, which governs the intolerant behaviour of fermions, is due to an underlying relativistic rotational symmetry, but that would stretch the point a little too far.

Other examples of forces not directly related to local gauge symmetries are the covalent chemical binding discussed in the preceding chapter, and the so-called `Van der Waals force` between atoms and molecules. The latter is the force that makes water cling to one's skin, or that prevents water from getting through the skin: like the covalent force, it is due to the quantum interaction of the outer electrons of atoms. Again, there is an underlying symmetry, related to the possibility that electrons may trade places, but this is not a gauge symmetry.

A local gauge twist causes wrinkles in the vacuum, because the mismatch between the twisted space and the unperturbed vacuum in the distance must be patched up somewhere in between. In the non-quantum picture, of which our tablecloth is a model, the difference is made good by wrinkles, 'field lines' that radiate from the twisted spot. But in a quantum world, where interaction is all-or-nothing, *the twist must be taken away by a single quantum*. The exchanged quantum, the velpon, then becomes the carrier of the vacuum wrinkle caused by the local gauge twist. According to what was said previously about Feynman paths and superposition, the net effect (on a large scale) of a gauge force is due to the superposition of all possible velpon exchanges that are allowed by the symmetry.

Since the vacuum tablecloth is actually four-dimensional space-time, it turns out that the velpon cannot patch up the mismatch in the vacuum unless it has at least four components. Thus, it must have spin 1 (or more) in order to have enough data-carrying capacity. With spin 1, the velpon is a *vector boson*.

10.5 The electromagnetic field

Let us now consider the simplest case of local gauge twisting. Since we anticipated that all gauge groups act like rotations, the simplest case is a rotation in a plane. Thus, we are looking for a single quantity (other than position in space and time) that is associated with a particle and that behaves like the angle of some rotation. Does such a quantity exist? The answer is yes, and you probably remember having seen this quite a few times: it is the *phase* of a particle. Suppose, then, that phase symmetry is a general property of Nature which – if applied locally – produces field wrinkles in the vacuum.

With this picture in mind, we can re-interpret our Feynman diagrams. For example, consider an electron moving through space. As we have seen, all quanta behave like waves. The direction and the wavelength determine the momentum vector of the quantum, and the frequency determines its energy. Momentum and energy can be observed directly, but the most characteristic wavelike property, the phase of the wave, cannot be determined by experiment. Only phase *differences* have observable consequences: they determine the interference behaviour (remember that the interference is what gave rise to the wave idea in the first place).

Because only phase differences matter, we should expect that the Universe is symmetric under global phase transformations. The reason is simply this: as we saw in the discussion about interference between alternative paths, the net probability is not changed if we rotate all the phase angles by a fixed amount. All that would happen is that the total amplitude vector would be rotated over the same angle, but since the probability is equal to the square of the length of the arrow, nothing observable would change. Thus, we can take the whole Universe by the scruff of the neck and give it a global phase rotation, and everything would keep running just as before.

But global symmetries cannot be applied because of relativity. Thus, our electron ought to be invariant under *local* phase changes. This is a local symmetry, and of course it cannot occur without a corresponding wrinkle in space-time. The quantum carrying the twist caused by the phase rotation

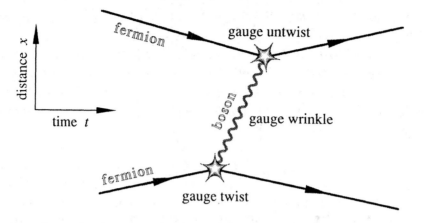

Fig. 10.5 First-order Feynman diagram of the scattering of two electrons. At the first vertex, a local gauge twist creates a velpon (field boson), which is annihilated at the second vertex. This exchange transfers momentum and energy between the two electrons.

is the photon, the particle of light. One possible Feynman diagram of Coulomb repulsion is therefore an exchange as shown in Fig. 10.5. Read this with your Feynman diagram scanner and note how the wrinkle in the vacuum, created by the local gauge twist, propagates to the other electron, where it untwists the vacuum by changing the phase of the other electron.

The phase of a wave is but a single number: the location of one wave crest on the path of the wave. To symbolize the dependence of the symmetry group on one number only, it is indicated by the symbol U(1). Because of the one-dimensional nature of the symmetry group, its velpon has a simple structure and carries a simple message, for instance '49 degrees': a single number, which suffices to specify to what extent the vacuum has been twisted out of kilter. But that's all; in particular, the single number does not contain enough information to transmit data about the strength of the coupling with the electron. Because this strength is measured by the charge of the electron, our simple velpon is uncharged.

Moreover, if the gauge symmetry is perfect and exact, then the wrinkles of the field extend indefinitely into space; because of the exactness of the symmetry, every remaining lack of smoothness, no matter how small, is noticeable, so that we have to keep patching up the mismatch between the twist and the rest of the universe, no matter how far away we go. Consequently, the U(1) velpon has infinite range, which, as we saw before, is a property that only massless particles have. Remember that the energy–time uncertainty relation reads $\Delta E \, \Delta t > \hbar$; because $E = mc^2$, this can be rewritten $c \, \Delta t > \hbar / c \, \Delta m$. In other words, if the mass of the particle is uncertain by an amount Δm, then the particle can be embezzled during the time Δt given by this relationship. Because $c \, \Delta t$ is the largest distance over which the particle can move during the embezzlement, the range of the particle is $c \, \Delta t$. But if we know that the mass is exactly zero, then it is not in the least uncertain, so that $\Delta m = 0$ and the range $c \, \Delta t$ is infinite.

Accordingly, the velpon corresponding to a U(1) gauge twist (merely a phase change), is the neutral, massless photon, which has spin 1. Because of the simplicity of the U(1) group, the photon is a comparatively simple thing; since it carries no charge, it cannot change the charge of the particle it is coupled to, nor can it contain enough information to carry any of its other properties. Hence, a U(1) velpon leaves the particle it is coupled to unchanged, except of course for its momentum and the orientation of its spin. Thus we see, at a photon vertex, an electron coming in and going out, coupled to a massless photon; and the same holds for the vertex of other electrically charged particles, such as protons or positrons. The theory of this process† is called `quantum electrodynamics`.

10.6 Playing with phase

For the benefit of people who are a bit unhappy about physics-by-analogy, I will expand a little on the way in which a local phase rotation generates an electromagnetic interaction. To begin with, let us go back to the discussion of the motion of quanta as derived by the superposition of Feynman paths. Classical mechanics was based on the law of inertia: 'A particle on which no force acts moves with constant velocity'. This rule was then extended to one for all classical motion: 'A particle moves along the path of stationary action'. The quantum analogue of this turned out to be: 'The most probable path of a particle is the path of stationary phase.'

We also saw how a phase change along a given path causes a deflection in the trajectory of a particle (for example, in Snell's law of refraction). Thus, in the Feynman picture, *a force corresponds to a prescription for changing the phase.* The particle 'finds' the path of stationary phase because interference makes that path the most likely, even though there are perfectly legitimate paths that go via Sirius. We found that

† See Feynman, *QED*.

(phase change) = $(1/\hbar)$ × (field density) × (step along path in space-time), or in symbolic form† $\Delta\phi = F\,\Delta x/\hbar$.

The goal now is to connect this phase-change prescription to the mechanism of a local symmetry transformation. First we note that if the field density F is constant all over space-time, the net phase change along all paths is the same. Therefore, a constant F will not produce a change in the path of largest probability. For example, if, in our discussion of Snell's law, we had changed the wavelength of the particle in all of space, we would not have produced a kink in the path. Only a space-time change of F is observable. It is *the difference of F from point to point in space-time* that produces changes in interference, and thereby changes the most probable path.

Of course, we, in our large-scale world, do not say 'there has been a change in this tennis ball's most probable path', but we say 'this tennis ball has been whopped with a racket': a force has acted on it. By the way, the fact that only the field *differences* count is inherent in the underlying phase mechanism: it is a consequence of the fact that phase is *periodic*. No matter how often the muddy phase wheel has gone around, the whole revolutions do not count; the little bit left over at the end is the only part of the phase that effectively contributes to the amplitude arrow. Thus, the wave behaviour of Nature, shown most clearly by the appearance of phase, expresses itself by the pervasive occurrence of *differences* in mechanics. When attending lectures on classical mechanics, I puzzled in frustration over the reason why a 'potential' had to be 'differentiated'. Now we see that it is the behaviour of phase that is responsible for the fact that, in classical mechanics, the equation of motion is a differential equation.

If we suppose that the phase change corresponding to the field density F is due to a global phase rotation (one that is the same over all of space-time), nothing happens. But a *local* phase change is, by definition, different from point to point. In this way, we make the connection with the local U(1) symmetry group. Algebraically, the symmetry is included by letting F be proportional to the amount of phase rotation, the amount of twisting of the cosmic tablecloth. Physically, F represents a field quantum, the famous 'second quantization' that identifies a wrinkle in the tablecloth with a boson created at the vertex where the twist occurs (Fig. 10.6). Writing out the

---

† In order to keep things relativistically correct, i.e. Lorentz invariant, we must take an infinitesimal step in space-time, not in space, so that Δx is really a little four-vector. In order to avoid cluttering the notation I will not write this out explicitly. Furthermore, the way in which the phase wheel leaves its prints when rolling through space-time (rather than through space) is more involved, but we will not need so much detail here.

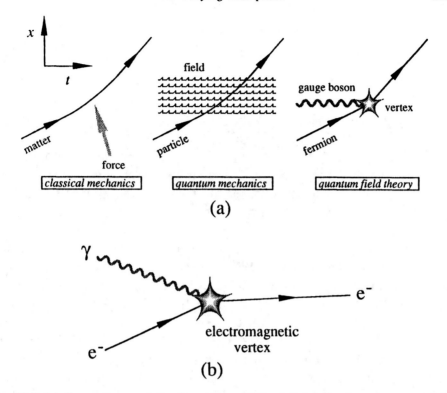

Fig. 10.6 (a) Successive quantization steps. Left, deflection of a matter particle by a force. Middle, representation of the force by a potential field. Right, the contemporary, fully quantized view: the force on a particle is a direct interaction with a field quantum. (b) The case where the interaction is electromagnetic: coupling with a photon at the interaction vertex.

algebraical details is a technically involved process, which I will not trot out here. But we can easily see one more relevant property of F, as follows.

Because Δx is a step along a path in space-time, it is a four-dimensional object. To continue the previous analogy, the muddy wheel traces out the wavelength of the particle in space, even as it traces the oscillation period in time. However, the phase is a scalar, a pure number, which is a one-dimensional object. Therefore, F can *not* be a scalar; we cannot say that a one-dimensional thing is a multiple of a four-dimensional thing (such as 'the number 42 is eight times a sphere'). So F must have several components; in particular, its number of components must be a power of four. This is how the four-dimensionality of space-time pervades the whole construction of fields and forces! Thus, the simplest possible quantum that can occur in F is a four-component beastie, a spin-1 boson like the photon.

By the way, in practice physicists use the expression $\Delta\phi = (e/\hbar)A\,\Delta x$ instead of the relationship $\Delta\phi = F\,\Delta x/\hbar$ introduced above. The field density A is called the relativistic `vector potential`, and the scalar quantity e is called the `electric charge`. We see that a change in the sign of e corresponds to a change in the sign of the phase. Therefore, $+e$ and $-e$ have opposite interference behaviour, as it were; in this way, the exchange of photons (as encoded in the field density A) can lead to *attraction or repulsion*. Explaining precisely how the exchange of a velpon can cause attraction is a bit of a tall order. The reason it is so complicated and so counter-intuitive is that this behaviour is basically due to the spin of the velpon. And spin, as we saw in the case of the spinning top, is *very* non-intuitive, even in classical mechanics.

A word of caution: it is important to distinguish between 'electric charge' and the generic word 'charge'! When we speak of electric charge, we mean specifically the quantity e introduced above, which is linked to the local U(1) symmetry that produces the electromagnetic field density A. In the more general case, charge is introduced in exactly the same way, namely as a proportionality constant in the phase shift, but the symmetry group under which F must be symmetric can be different. Thus, we have weak charge associated with the gauge group SU(2), and colour charge in the case of SU(3). We will discuss these cases below.

In summary, the electromagnetic behaviour of the electron can be constructed as follows. First, we note that the world should not change under global rotations of the quantum phase: nothing moves differently if we rotate the universal tablecloth as a whole. The discovery of a symmetry among the members of a fermion multiplet (here, simply the singlet of the lone electron) suggests that we make this a local symmetry, because the laws of relativity do not allow instantaneous action at a distance. Second, a change in quantum phase produces a deflection of the path of largest probability. Third, suppose that the electron is symmetric under the phase rotation group U(1). Then a local U(1) twist changes the phase of the electron, and thereby deflects its path. Fourth, there must be a quantum carrier of the U(1) twist; it is called the photon. Fifth, the electromagnetic interaction of other particles (e.g. the proton) should proceed in precisely the same way.

10.7 Isospin

In the case of quantum electrodynamics, the phase change due to a local gauge twist can be written as (phase change) = $(1/\hbar)$ × (electric charge) × (photon field) × (space-time step). This is the essence of the Feynman approach: every unit of phase rotation corresponds to $\hbar$ units of action. Such a picture immediately suggests that we try to describe other forces by means of the prescription (phase change) = $(1/\hbar)$ × (charge) × (velpon field) × (space-time step). The velpon field would be generated by some local symmetry, and the quantity called charge would indicate how much phase change every velpon produces. Thus, the charge is a measure of the importance of the velpon, as compared to the velpons produced by other symmetries; the charge is a `coupling constant`.

With this goal in mind, our program would be (1) identify suitable fermion multiplets, (2) construct the symmetry group that is supposed to turn one multiplet member into another, (3) assume that this is a local symmetry and derive the properties of its velpon field, (4) calculate the interactions of the multiplet particles and see if they fit their observed behaviour, (5) if possible, observe the velpons directly.

To illustrate this approach, I will first present an example that is suggestive but that does not actually work. Historically, things did go approximately this way; moreover, the example is simple enough to illustrate the main points of the chapters that follow.

A change in phase is described by a single number, and hence the electromagnetic quantum, the photon, is a simple beast. To begin with, the gauge symmetry that creates it (phase rotation) is exact, so the photon is massless. That an exact local symmetry creates a massless velpon can be glimpsed as follows. Suppose that we make the local twist of our tablecloth smaller and smaller. Because the phase change is proportional to the amount of twist (= the field strength F), the phase change goes to zero as the wrinkling disappears. However, a massive velpon always has a *finite* field strength: if it has mass m, then its energy can never be smaller than the relativistic value mc^2, so that, when it interacts, it delivers a full wallop (and remember, quantum interaction is all-or-nothing, so that a 'careful' interaction is impossible). A massive particle cannot be turned off gracefully, as it were, whereas in a massless particle (such as the photon) the momentum and energy vanish *simultaneously* when we let its wavelength go to infinity.

The photon is also simple because the phase change is described by a single number, so that phase changes produced by photons are commutative: the order in which they are executed does not matter, so the photon gauge group U(1) is Abelian. Therefore, the photon has no hidden memory of the way in which it was created by the gauge twist. Thus, the photon carries no information about the strength of the coupling at the vertex, and, consequently, the photon carries no electric charge.

Occasionally, one finds sets of fermions that resemble each other very closely in some respects. One such set is the proton–neutron pair (p, n). In atomic nuclei the protons do not fly apart, so there must be an attractive force between them that can overcome the electric repulsion among the protons. Because all multiple-proton nuclei also contain neutrons, it is natural to suspect that this force is an attraction that acts between protons and neutrons. Experiments show that this force, the `strong force`, exists and does indeed keep the protons and the neutrons together. Moreover, the force does not only act between a proton and a neutron, but also between two protons or two neutrons.

In fact, it turns out that in strong-force experiments the force between p and n is practically the same as that between p and p or between n and n. The similarity is also apparent in the fact that the light stable nuclei, from deuterium (2 nucleons) up to about sulphur (32 nucleons), contain exactly or very nearly equal numbers of protons and neutrons. This almost identical behaviour with respect to the strong force, together with the fact that the proton's mass is only 0.14% less than that of the neutron, indicates that the physical similarity of these particles is much greater than mere chance would allow.

Consequently, we are led to suspect that, in some sense, the neutron and the proton are different guises of the same particle (called `nucleon`): in one state it is a neutron, in the other state it is a proton. This symmetry is called `isospin symmetry`. In quantum language, we would say that the nucleon is a superposition of proton and neutron. Suppose we call the amplitude of the nucleon N, and let p and n be the amplitudes of the proton and the neutron. Then we can write† their superposition as $N = p \sin \theta/2 + n \cos \theta/2$, as in the case of the two-component chimaera (lion mixed with goat).

The angle θ indicates the amount of 'protonness' and 'neutronness' present in the nucleon (Fig. 10.7). By changing that angle, we can rotate the nucleon in isospin space such that it changes from being (say) a proton to being a neutron. This is a symmetry operation called an `isospin rotation`. Imagine that we were to subject the entire Universe to an isospin rotation, in such a way that everywhere there was a proton there is now a neutron, and vice versa. If isospin symmetry were exact, this operation would leave the world running exactly as it was, and we would conclude that the Universe is invariant under a global isospin group. We *observe* that the proton and the

---

† There is a technical difference, in that we use a *half-angle* formula, containing $\theta/2$. The reason for this is that the probability P is equal to the *square* of the amplitude N. Now $\cos(0) = -\cos(180°)$ and $P = N^2$, so that $\theta = 0$ and $\theta = 360°$ describe the same particle only if we use the half-angle in the superposition.

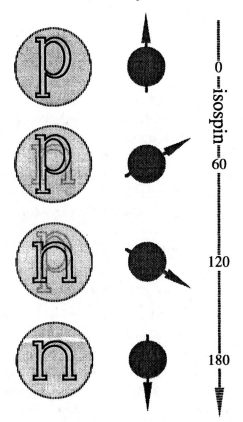

Fig. 10.7 Illustration of the nucleon as a function of the mixing angle. The top panel shows a pure quantum state (the proton), which is defined as isospin-up. The bottom panel shows another pure state of the nucleon: the neutron, corresponding to isospin-down. The other states are superpositions in between those extremes.

neutron are, in many respects, astonishingly similar, so we must take isospin symmetry seriously, even though we see immediately that isospin symmetry is *not* exact: the proton is electrically charged but the neutron is not, and their masses are not the same.

The quantum laws imply that if two states can be superposed at all, they can be superposed with an arbitrary mixing angle; thus, an isospin transformation is a smooth rotation-like symmetry (not discrete, such as the head-or-tail of a coin), albeit a rotation in isospin space and not in space-time.

Because we know of no other particles that resemble the proton and the neutron nearly as closely as these two resemble each other, we presume that the nucleon has precisely two states: it is a doublet under the isospin group. Accordingly, the isospin group must be two-dimensional. In this respect, isospin symmetry is more complex than the U(1) gauge symmetry that created the photon. Whereas U(1) is Abelian (so that the photon carries no charge), *the isospin group is non-Abelian*, like the spatial rotations of dice.

The order in which two successive non-Abelian twists are performed is important: if A and B are isospin twists, then in general $A \otimes B$ does not equal $B \otimes A$. We saw earlier that two successive Lorentz boosts† cannot in general be replaced by a single Lorentz boost, but instead is composed of a Lorentz boost plus a spatial rotation. Similarly, non-Abelian symmetries multiply as $A \otimes B = C \otimes (?)$, in which the question mark stands for an internal degree of freedom of the velpon, and A, B, C are symmetry rotations. As it happens, the extra degree of freedom indicated by (?) is associated with the charge of the velpon: *non-Abelian velpons can carry charge.*

10.8 Yang–Mills fields

Let us ignore differences of mass and charge between p and n, and assume that isospin symmetry is a genuine symmetry of Nature. Thus, if electromagnetic effects (which mostly act on the proton) are negligible, the nucleon is invariant under isospin rotations, and it is immaterial whether we deal with neutrons or with protons, or with some quantum chimaera in between. The fact that all signals take time to cross space makes a global symmetry impossible, because we cannot execute an isospin rotation everywhere in the Universe at once. Hence we must consider local isospin symmetry. (Besides, a truly exact global symmetry operation would have no observable consequences anyway: compare (a) of Fig. 10.3, showing a global rotation, with (b), showing a local twist and the wrinkles it causes.) The tablecloth experiment tells us what to expect: local symmetry can exist only in conjunction with a field of wrinkles that patches up the difference between the locally twisted vacuum and the global Universe at great distances.

In the case of the U(1) group which generates the photon, we saw that a local phase rotation produces a field by the process $e^- \rightarrow$ (local U(1) twist) $\rightarrow$ $e^- + \gamma$, where γ stands for the photon. In the case of the nucleon, we have four possible cases because the nucleon is not a singlet like the electron but

---

† A Lorentz boost is a Lorentz transformation that keeps the direction of the spatial coordinates the same.

a doublet:

$$p \rightarrow \text{(local isospin twist)} \rightarrow p + \text{velpon type 1}$$
$$p \rightarrow \text{(local isospin twist)} \rightarrow n + \text{velpon type 2}$$
$$n \rightarrow \text{(local isospin twist)} \rightarrow p + \text{velpon type 3}$$
$$n \rightarrow \text{(local isospin twist)} \rightarrow n + \text{velpon type 4}$$

The transmutation $n \rightarrow n$ can be produced by $n \rightarrow p$ followed by $p \rightarrow n$. If we reverse the order we get a similar do-nothing reaction $p \rightarrow p$ which may change the momentum of the particle but not its identity. Thus, the type-4 reaction and the type-1 can be caused by the same velpon, so that the local isospin twist produces *three* different field quanta. This turns out to be a general property: *if an N–plet of fermions interacts through a local symmetry, then there are $N^2 - 1$ velpons involved in the interaction*. Notice the predictive power of this type of theory: it severely restricts the number and properties of the velpons. For every N, there are $N^2 - 1$ massless vector bosons.

The field that is created by local isospin rotation is called the `Yang-Mills field`. Its velpons are more complex than the photon because the gauge twist corresponding to the isospin symmetry cannot be described by a single number: isospin symmetry is a flip between *two* states. The isospin group is called the `special unitary group of dimension 2`, or SU(2).

The SU(2) group is non-Abelian, so that the order in which successive gauge twists are executed is important, and the velpons have a hidden memory of the strength of the interaction at the vertex. This means that the Yang–Mills velpon need not be neutral, in contrast to the photon: it can carry 'isospin charge'. Because the neutron and the proton have different electric charges, the Y–M velpons must carry electric charge as well. One of the most important consequences is that the Yang–Mills velpons can *interact directly amongst themselves*, something that the neutral photon is incapable of doing (Fig. 10.8). Note that all this richness is due to the fact that the SU(2) group is non-Abelian.

Is there a set of three particles in Nature that could be identified with the Yang–Mills velpon? It seems that there is. There exists a particle called the π-meson, or `pion`, that comes in three guises: a positively charged π^+, a neutral π^0 and a negative π^-. By analogy with the electromagnetic case, we expect the vertices drawn in Fig. 10.8, and equivalents in which the time order of events has been changed by a Lorentz transformation, as in Fig. 10.9. The strong force is then described by Feynman diagrams of the type shown in Fig. 10.10.

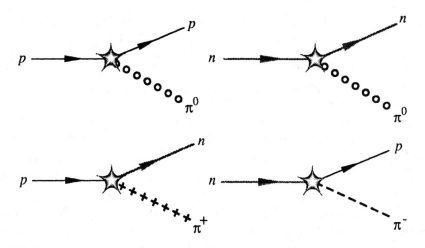

Fig. 10.8 Proton–neutron–pion vertices. The proton has one positive electric charge, which is carried away by the π^+; the neutron and the π^0 are electrically neutral.

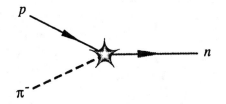

Fig. 10.9 A proton and a negative pion combine to form a neutron. This diagram can be obtained from the second one on the left of Fig. 10.8 by inverting the time order of the positive pion.

Note that *the type of nucleon can change at a vertex*, for example from proton to neutron if a π^+ is emitted or a π^- is absorbed. This never happened at the electromagnetic vertex: an electron stays an electron in electromagnetic interactions, and so does a proton and all other electrically charged particles. The magical change comes about because SU(2) is non-Abelian, whereas U(1) is Abelian; therefore, the Yang–Mills velpon can carry 'isospin charge' and can change the iso-identity of a particle it interacts with.

The pion triplet seems to be a brilliant candidate for the role of the isospin velpons. They form an isotopic-charge and electric-charge triplet (π^+, π^0, π^-). Their masses, about 275 electron masses for π^{+-} and 268 for π^0, imply an energy embezzlement lifetime of about 10^{-23} seconds and a corresponding range of flight of some 10^{-15} metres. Since this is the typical

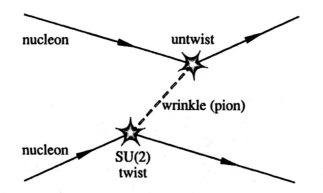

Fig. 10.10 First-order Feynman diagram of nucleon–nucleon scattering. This is completely analogous to the electron–electron interaction diagram shown earlier. However, it will be seen that this interaction does not obey the rules for forces that are due to local symmetries, and therefore cannot describe a fundamental force.

distance between protons and neutrons in atomic nuclei, we may be sure that the pions produce the strong force.

Everything seems fine so far, until we remember that the exact local U(1) symmetry that creates the photon also ensures that the photon is massless. Because we had assumed exact local SU(2) symmetry, and because velpons are vector particles, we expect the pions to be massless and have spin 1. But when these properties are measured in actual pions, it turns out that pions *do* have mass and do *not* have spin. Moreover, they are not stable; the decay time of the π^0 is 10^{-16} seconds, which seems to be no cause for panic because it is a comfortable ten million times longer than the timescale of strong interactions, but the π^{+-} live for 10^{-8} seconds, a whopping hundred million times longer than their neutral buddy π^0! Why? Is the Yang–Mills field a mirage after all? In the next chapter, we will return to this point. But since we are discussing the consequences of all possible local symmetries, I want to make some space here for the odd one out among gauge fields: gravity.

10.9 Gravity, a splendid failure

In the presentation above, we have always used the Lorentz transformation in a global way, applying the same symmetry all over space-time at once. Yet the fact that the speed of light is finite and maximal induced us to drop global symmetries, and to use local symmetries instead. We saw that the mismatch between a locally twisted patch of space-time and the rest of the world can be made up by a field of wrinkles, which appear on a large scale as the field

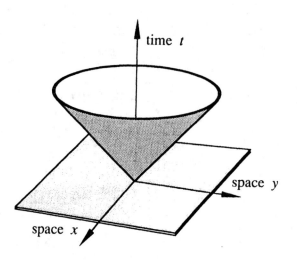

Fig. 10.11 Two-dimensional light cone. This diagram is obtained by stacking the expanding light wave along the direction of the time axis. The light wave expands like a ripple on a pond. At any time we can find out how far the light wave has travelled by cutting the cone with a plane perpendicular to the time axis. The resulting circle shows how far the wave has advanced.

lines of a force. We ought to treat the Lorentz group like the rest, and use it as a local symmetry. The force transmitted by the wrinkles generated by a local Lorentz transformation has a name: we call it gravity.

Historically, gravity was the first force to be described by means of a local gauge symmetry. How that comes about can be seen as follows. We recall that a rotation in space leaves the Pythagoras distance $x^2 + y^2$ constant, and that a Lorentz superrotation in space-time keeps the Minkowski distance $r^2 - c^2t^2$ invariant. Thus, the lines $r = \pm ct$ are the same in all coordinate systems. For example, if some observer is moving with constant speed v with respect to space-time as described by the (r, t) coordinates, then that observer has different coordinates, say (R, T); but one always finds that the lines $r = \pm ct$ correspond to $R = \pm cT$. These lines are the paths along which light travels from the point $r = 0$, $t = 0$. They define the light cone, so called because the expression $x^2 + y^2 + z^2 - c^2t^2 = 0$ describes a three-dimensional cone around the t-axis in space-time. Suppression of the z-coordinate gives $x^2 + y^2 - c^2t^2 = 0$, a cone that is easier to draw or to make out of paper (Fig. 10.11).

Why is the light cone useful? Simply because it is invariant under Lorentz rotations. Thus, the light cone is a good compass for navigating in space-time, and for indicating its structure. For example, a global Lorentz rotation

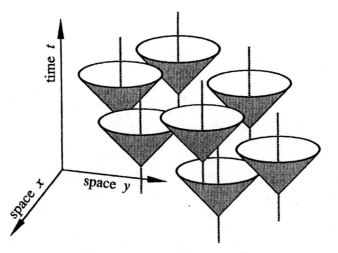

Fig. 10.12 Light cones in flat space-time. All the light cones are neatly arrayed; there is no mismatch between nearest neighbours. This type of space-time is called flat. The space-time is globally symmetric under Lorentz transformations.

does not change the relative orientation of the light cones anywhere in space-time. Thus, in a globally Lorentz symmetric world we can choose our coordinates such that all light cones point in the same direction everywhere (Fig. 10.12). By means of the light cone we can show, in a graphically obvious way, the warping of space-time that corresponds to the wrinkling of the vacuum tablecloth.

What if Lorentz rotation is used as a local symmetry? In that case, the amount by which space-time is twisted differs from point to point and from moment to moment. A given light cone need no longer point in the direction of its neighbours. Locally, of course, nothing happens: if I choose a small enough patch of space-time, the shape of the light cone remains the same, and the orientation of closely adjacent cones† is very nearly equal. But on a global scale, things may begin to look strange. This is, of course, what we expected all along: a local symmetry causes wrinkles in the vacuum, and *a local Lorentz transformation causes wrinkles in space-time.*

The appearance of warps in space-time can be seen quite nicely if we connect the coordinate systems of nearest-neighbour light cones (in pictures, this is easiest if we suppress all but one space dimension; see Fig. 10.14).

---

† This is what appears in some textbooks as 'a locally free-falling coordinate system'. Saying that in a small enough patch of space-time the light cones are closely parallel is the same as saying that it is always possible to choose a reference frame where space-time is locally flat; this flatness is then identified with the absence of gravitational acceleration.

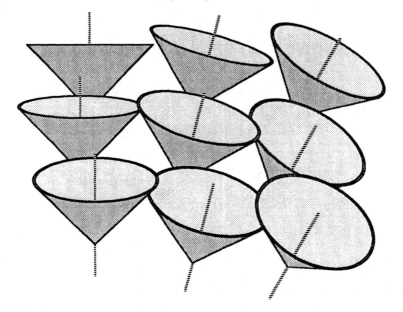

Fig. 10.13 Light cones in arbitrarily curved space-time. Now the light cones are no longer neatly arrayed: they are misaligned, so that there is a mismatch between nearest neighbours. This type of space-time is not flat, because it is not globally symmetric under Lorentz transformations. It is always *locally* symmetric, though: the shape of the light cone is still the same everywhere. Only its orientation varies from event to event.

To do this, we connect the light cones with lines that go precisely along their axes (i.e. the local direction of time). The imaginary thread that runs through the light cones, tying them together like a string of conical beads, is called a `free-fall line`.

Similarly, we can connect the light cones with lines that run precisely perpendicular to their axes (in three dimensions, these lines correspond to small bits of space, but that is a little difficult to draw!). The resulting lines, from which the light cones protrude like pins in a pincushion, are called `equal-time spaces`. By construction, the free-fall lines and equal-time spaces are always mutually perpendicular. But in general they do not form a regular rectangular grid. If they do, we have a special case called `flat space-time`. Otherwise, they are crooked: *the wrinkles caused by a local Lorentz symmetry appear as a warping of space-time.* This is the origin of the famous `curvature of space-time` that you may have heard about: the sewing together of oddly oriented pieces of space-time can produce a cloth with a complicated geometrical shape.

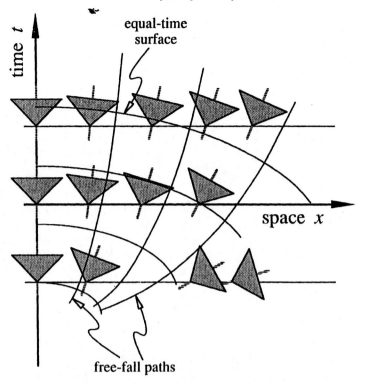

Fig. 10.14 Light cones in a small section of Friedmann space-time. An arrangement of light cones such as this occurs in our Universe on a very large scale. The distance to which we can look back is equal to the age of the Universe, about 15 billion light years; our Galaxy has a size of some 80 000 light years, two hundred thousand times smaller. Because the Friedmann space-time is spherically symmetric, every observer in the Universe sees an arrangement like this.

How would we observe that curvature? Let us consider the difference between a straight and a curved track in space-time. If the track is straight, equal distance intervals Δx are covered in equal time steps Δt: the velocity $\Delta x/\Delta t$ is constant. But if the track is curved, then different space intervals are covered when time increases by Δt. The velocity is then not constant: we have an *acceleration* (Fig. 10.15) with respect to an observer at the origin of the coordinate system. A particle moving along a free-fall path is blissfully ignorant of any acceleration as long as it just minds its own local business. But if the particle looks around, it sees that its neighbours change their relative velocities. Thus, the curvature of space-time can be interpreted as an accelerating force: `gravity`.

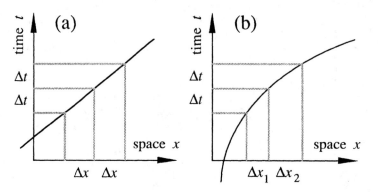

Fig. 10.15 Uniform and accelerated motion in a space-time plot. (a) Uniform motion has a constant velocity. In equal time intervals Δt equal distances Δx are traversed. (b) When the amount of space Δx_1 covered in the first time step Δt is not equal to the amount Δx_2 traversed in the second step, the motion is no longer uniform but accelerated. The space-time path is then not straight but curved.

It was Descartes who first pointed out that space has real physical attributes, namely height–width–depth (three dimensions), and that we are therefore justified in considering space as real stuff, of which our Universe is made. It is then perfectly natural to ask: 'What is the structure of space?', just as we are accustomed to ask about matter. It was Einstein's insight that the structure of space can be described completely by a prescription for the distances in it (or, more generally, space-time). The description of the geometry of space-time is called the general theory of relativity or GRT. In the earlier chapters we restricted ourselves to special relativity, which applies only to bodies moving with constant velocity relative to each other. We have now extended this to the general case of accelerated objects. The GRT describes the interactions of gravitating masses by means of space-time curvature, caused by local Lorentz rotations. In GRT parlance, we speak of 'a space-time dimple'; in poetry, the same thing may be called 'the Moon'.

How can a distance indicate warping? Consider the following: an archaeologist digs up three Sumerian clay tablets (Fig. 10.16). On one is written: 'A journey from A to B takes three days by camel.' On the second, it says: 'From B to C takes four days by camel.' And on the third: 'From C to A takes 11 days by camel.' Well, there is no plane triangle with sides of 3, 4 and 11! Is anything wrong with the inscriptions? No: as it happens, there is a mountain range between cities A and C! Cases like this one do, actually, occur in archaeology: a description of distances gives information about

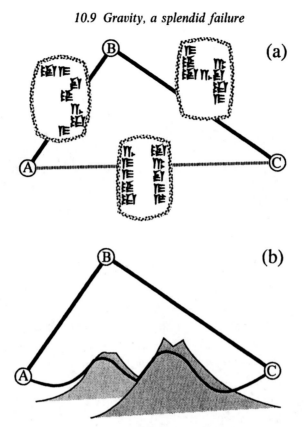

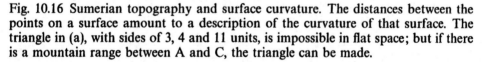

Fig. 10.16 Sumerian topography and surface curvature. The distances between the points on a surface amount to a description of the curvature of that surface. The triangle in (a), with sides of 3, 4 and 11 units, is impossible in flat space; but if there is a mountain range between A and C, the triangle can be made.

the surface along which the distances are measured (of course, the distance recipe is rather involved in this case).

Of course, you already knew that a distance recipe is the same as a curvature. If you have ever looked closely at the distance tables in an atlas, you will have noticed that such a table cannot be mapped on a plane piece of paper. If, instead of memorizing the table, you set about making a map of the distances, you would soon find that it cannot be done (Fig. 10.17). You would have to shrink the edges of your paper, or make cuts in it and paste the bits together, and soon you would find that your map has become a sphere: a correct map of planet Earth! I have tried it with the tables in my atlas and I obtain the radius of Earth correct to about 10%.

Just for fun, let us consider one special case of patching light cones together. Suppose that the geometry of space-time is independent of time,

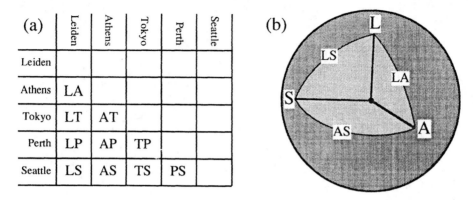

Fig. 10.17 The distance table (a) in an atlas cannot be mapped correctly on a flat piece of paper. Instead, if you carefully follow the distance recipe, you will end up constructing a curved surface (b): the sphere of Earth.

in the sense that the orientations of the light cones are the same on surfaces $t =$ constant. Next, suppose that the spatial part of space-time is spherically symmetric. Then only one space direction, the radial coordinate r, suffices – together with the time t – to describe space-time. In particular, the relative orientations of the light cones depend on r only. Now let the difference $\Delta\alpha$ in angle between the axis of one light cone and that of its neighbour, a distance Δr away, be given by $\Delta\alpha = \Delta r/(2r^2 - 2r + 1)$.

Clearly, the light cones tip over more and more as the radius r approaches zero (Fig. 10.18); there is a point where the light cones lie flat on their sides, namely at $r = 1$. When we study this curious pattern carefully, we realize that within the radius $r = 1$, all light cones lie entirely within a sphere with radius 1. Therefore, *from within that sphere no light can escape*: the surface $r = 1$ allows time-like paths to pass into it, but nothing can get out unless it moves faster than light, which is impossible because of Lorentz symmetry. The surface with radius 1 is called a `horizon`, and the entire space-time structure is known as a `black hole`. At large distances, we see that $\Delta\alpha \to 0$: the influence of the black hole vanishes when we move away from it.

However, none of this includes quantization. Although gravity is clearly a gauge field generated by the Lorentz group, it is purely a large-scale, classical field. The coupling between the gravitational field and its material source is established by linking the distance recipe in space-time (called the `metric tensor`) to the presence of matter, which must be inserted from the outside: it is not directly demanded by the theory. The coupling strength, or gravitational charge, is called `mass`; for historical reasons, this

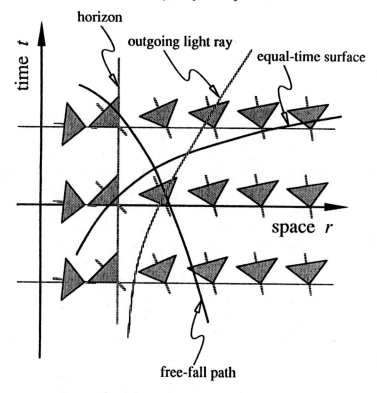

Fig. 10.18 Light cones of the Schwarzschild black hole. The vertical axis shows the time and the horizontal axis shows the distance as seen by an observer at infinity. At large distances from the time axis at $r = 0$ (called singularity) the light cones are almost exactly aligned. But the closer one comes to the singularity, the more the light cones tip over. The cones ultimately keel over so much that they lie flat on their sides. Thus, they enclose a region from within which no light can escape: a horizon surface. Space-time is so tightly curved that light cannot get out of this enclosure. Because nothing can go faster than light, nothing can escape at all: the surface is perfectly black.

coupling is written $M\sqrt{G}$, where G is the gravitational constant. In a quantum theory, this coupling would be established by linking a gravitating particle to a velpon, the graviton. The coupling strength $M\sqrt{G}$ at the vertex then connects the graviton field to the quantum phase shift, just as the electric charge e was used in the expression (phase shift) $= e \times$ (electromagnetic field) $\times \cdots$ before.

This course of action has been attempted, and even though Einstein's theory is a resounding success when applied to classical fields on a large scale, its quantum version has failed miserably. We can gain some idea why if we consider the behaviour of the coupling strength. If two particles of equal mass M interact via gravity, the force between them is proportional to GM^2. The analogous quantity in Coulomb's law is the square of the electric charge, e^2. The occurrence of the factor M^2 in the gravitational coupling spells its doom. Whereas the electric charge is a parameter that occurs unchanged throughout the theory, the value of M^2 depends on circumstances because the mass has the bad fortune to be present in the relation $E = Mc^2$. Every energy is associated with an equivalent mass, and therefore the coupling strength $M\sqrt{G}$ depends on the energy of the process it describes. Moreover, the coupling strength of gravity depends on the energy of the interaction in an unmanageable way. If it were $\sqrt{G}/M$ things would be safe, but with M in the numerator we get a runaway effect: the energy increases the coupling, which increases the severity of the interaction, which increases the energy and so forth. This destroys all hope of obtaining a well-behaved quantum theory of gravity along the lines we have followed until now.

   This situation is absolutely unacceptable, but no solution has been found. Maybe the bad behaviour of the coupling constant is only apparent, but at present the case against a straightforward type of quantum gravity seems pretty ironclad. More likely is that other basic interactions in Nature will be found to interfere destructively with the badly behaved processes in quantum gravity, thereby neutralizing the unruly behaviour of the coupling strength. But despite many determined attempts, no working theories of this type have yet been discovered for the gravitational field. One of the problems is that the experimental situation is so bleak. The electrostatic force between two electrons is 10^{37} times stronger than the gravitational force between them. Under these circumstances, it is likely that any effect of quantum gravity accessible with present-day particle accelerators would be hopelessly swamped by the other forces of Nature.

11

Colour me red, green and blue

☆

11.1 The colour field

The gauge symmetry explanation of the strong force ended in frustration: so near, and yet so far! If only those pions hadn't taken it into their little heads to be massive spinless particles, instead of being vector bosons as we would have liked. But maybe not all is lost; the isospin symmetry is unmistakable, because the similarities between the neutron and the proton are readily observable, and the pion triplet does smell strongly of SU(2). Maybe we can keep most of this, and just allow the possibility that the pion is a scalar particle because it is *composite*: if it consists of a spin up–spin down pair, it can have total spin zero. Also, if the pions are composites, it is no longer unexpected that they decay, even though the phenomenal difference between the π^{+-} and π^0 lifetimes needs an explanation. Maybe there is more structure in the strong force than meets the eye.

If the pions are composites, then we must assume that the nucleons are composites, too. Collectively, the pions, nucleons and their relatives – which we will meet anon – are called hadrons. What are the quanta of which the hadrons are made? In other words, what is the fermion multiplet from which we must build them? Consider an analogy with the local U(1) symmetry that produced the photon, electromagnetism and electrically bound composites such as atoms. First, we find a suitable N-plet of fermions; then their gauge group, acting as a local symmetry, produces $N^2 - 1$ velpons, analogous to the photon; these velpons bind the fermions together to form hadrons, analogous to the way in which electromagnetism binds electrons to atomic nuclei.

Immediately, formidable problems arise. To begin with, why have we not seen the constituents of the hadrons before? After all, it is not difficult to split an electron off an atom: a television picture is written with free electrons. Why can't we split hadrons similarly? And what is the value of N? As it happens, the answer to the second question is easier to guess than it is to answer the first (I will return to that in Chapter 13). We cannot very well have $N = 1$, or else we would obtain electromagnetism, which cannot produce a composite electrically charged particle as compact as the proton. Could we have $N = 2$? Not easily, because the proton and the neutron have spin $\frac{1}{2}$ and the two particles of a doublet would only produce bosons as bound states.

Thus it seems that $N = 3$ is the simplest possible case that does not contradict known particle properties. To our great good fortune, this appears to work. The fermions of the basic triplet are called `quarks` and the corresponding symmetry group is the special unitary group of dimension three, or SU(3).

In U(1), we had only one type of charge, the electric charge; conventionally, this is called −, or minus one unit, while its anticharge is called +. Because SU(3) is three-dimensional, there are three types of charge (as well as their anticharges). These types are called R, G and B while the corresponding anticharges are $\bar{R}, \bar{G}, \bar{B}$. The SU(3) charge has been given the whimsical name `colour charge`. The quarks then come in three 'colours': red, green and blue (complemented by antired, antigreen and antiblue). This has absolutely nothing to do with visual perception; it's just a metaphor, but − as we will see − a useful one.

The names of the colours were taken from the three basic phosphors in a colour television picture tube. With a magnifying glass, you can see that a white patch on the screen really consists of an assembly of bright red, green and blue dots. If you want names that do not suggest physical properties that might be confusing, the possibilities are endless: Huey, Louie and Dewey; or Faith, Hope and Charity; or whatever, as long as you don't think that quarks are actually coloured, ducks or charitable. In a hypothetical SU(4) theory, there would be four charges, which might be called north, east, south and west; or hearts, clubs, diamonds and spades; or whatever foursome suits the local culture.

As before, we calculate that a local SU(3) gauge twist must be matched to the unperturbed vacuum at large distances, which is possible if the corresponding field is composed of $3 \times 3 - 1 = 8$ velpons, each of which is a massless vector boson. With three colours, there would be nine gluons, but since one of these can be built up from the other eight, only $N^2 - 1 = 8$ gluons are meaningful velpons. You may want to spell this out for yourself in analogy with the isospin transformations between the proton and the neutron. If we call the three quarks (q_R, q_G, q_B), we have $q_R \to$ (local SU(3) twist) $\rightarrow q_R +$ velpon type $R\bar{R}$; $q_R \rightarrow q_G +$ velpon $R\bar{G}$; and so forth with other colour combinations. In electromagnetic pair creation, we saw that adding electric charge (e.g. an amount $-e$ for an electron) has the same consequences as removing an anticharge (such as $+e$ for a positron) on the total balance sheet of electric charges. Likewise, adding a colour is the same as removing an anticolour, which is the reason the velpons carry one colour and one anticolour (for example $R\bar{G}$ or $R\bar{R}$). As before, the ninth velpon can be written as a combination of the other eight, and so can be dispensed with, but this cannot be done in a very transparent way.

Since SU(3) is non-Abelian, the velpons (which in this case are called gluons) can interact with one another because some of them carry the charge corresponding to the SU(3) symmetry. Hence, the emission of a velpon can change the colour of the quark, just like the emission of an isospin particle (e.g. a π^+-meson) can change the iso-charge of the nucleon to which it is coupled (e.g. from p to n when a π^+ is exchanged).

The basic multiplet of particles in the electromagnetic gauge group U(1) is a singlet, i.e. it contains one fermion (such as an electron, a positron, a proton and so forth). The basic multiplet of SU(3) is a quark triplet of three different colours, which have been labelled red (R), green (G) and blue (B). If a quark emits a gluon, it may change colour; for example, a red quark can become a blue quark by shedding a gluon that carries off the red and adds the blue (Fig. 11.1). In the language of quantum mechanics, adding the blue charge is the same as removing the antiblue; therefore, a red quark can be turned into a blue quark by a local SU(3) twist that emits a red–antiblue ($R\bar{B}$) gluon.

In the colour metaphor, there is an elegant way to summarize the fact that the velpons patch up the damage done by a local gauge twist: *gluons are exchanged to paint the world white*. The gluons zip from event to event, thereby cancelling local colour differences between quarks in such a way that there is always a full set of complementary colours present: red, green and blue; or red and antired, and likewise for the other colours.

Herein we observe an analogy with the electromagnetic force, generated by U(1). In that case, we could have called positive electric charge black and negative white, and we could have said that 'photons are exchanged to paint the world grey'. The exchange of photons that bind the electron to the proton in a hydrogen atom produces an electrically neutral state (an electric singlet). For historical reasons, we say 'opposite electrical charges attract each other'. In the case of the colour force, we say 'coloured particles assemble to form white composites'. In a white composite (a colour singlet) it does not matter, as far as the net colour is concerned, if I change all colours:

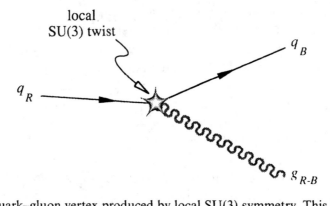

Fig. 11.1 Quark–gluon vertex produced by local SU(3) symmetry. This construction is completely analogous to the electron–photon vertex of the quantum electrodynamic theory which describes electromagnetism. Accordingly, the colour SU(3) theory is called quantum chromodynamics. In the example shown here, a red quark q_R changes colour and becomes a blue quark q_B. The colour difference $R - B$ is carried away by the gluon g_{R-B}. The gluon can carry colour charge because of the complicated non-Abelian structure of the SU(3) group, which governs the local symmetry twist at the vertex.

replacing $R \to B$, $B \to G$ and $G \to R$ keeps a white particle unchanged, no matter how it is built inside. In other words, a white particle behaves with respect to rotations in colour space in the same way as a sphere behaves under spatial rotations: it does not change. A white state is a singlet under the colour symmetry.

Because there are three colours, we expect that the quarks form white composites in at least two different ways: first, by combining a quark of a given colour with an antiquark of the corresponding anticolour; second, by combining three quarks of three different colours. These combinations are the simplest white quark composites, and just as the simplest electrically neutral atomic composites (e.g. one proton plus one electron = a hydrogen atom) are the most plentiful in the Universe, so we expect the two- and three-quark combinations to be the most common. These combinations are the hadrons (Fig. 11.2).

And common they are: the quark–antiquark composites are the mesons, of which the three pions are examples, and the three-quark composites are the baryons, of which the proton and the neutron are examples. In this picture, composite bosons (mesons) are exchanged between composite fermions (baryons), an exchange that appears outwardly as the strong

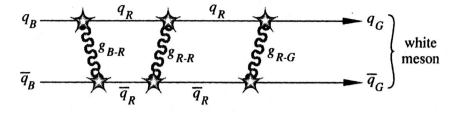

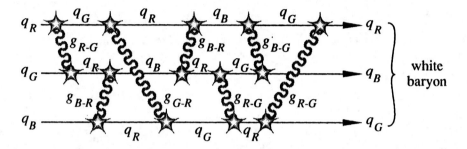

Fig. 11.2 Diagram of how mesons and baryons are glued together: the constant exchange of gluons creates a permanent force field which binds quarks tightly to each other. Such bound states (colour singlets) are colour-neutral or white. This is similar to the exchange of photons between electrons and protons, which keeps atoms together: an atom in equilibrium (electric singlet) is electrically neutral.

`force`. Hence the colour force appears in disguise, and under the pseudonym of the strong force, which (as we saw in our earlier SU(2) attempt) is not a gauge force. That is clearly shown in the fact that composites are exchanged here, and not single velpons. The colour force proper is not shown in Fig. 10.10; but it is there, and it serves to hold the baryons and the mesons together (Fig. 11.3).

11.2 Strawberry fields forever

There is no known reason why the story of the colour force should not stop here, but things are not that simple. Indeed, the attentive reader will already have noticed that the prescription 'a baryon is a white composite of three quarks' seems to allow the existence of only one baryon, namely $q_R + q_G + q_B$ and, of course, its antiparticle. Using antiquarks does not help, because an anticolour is equivalent to the sum of two other colours: if red + antired =

white, and if red + green + blue = white, then in some sense antired = green + blue. Consequently, if we start building a white baryon beginning with an antired quark, we have no choice but to add an antigreen and an antiblue one, thereby ending up with the antiparticle of the baryon we had already. This is blatantly contrary to observations. We have, even in our everyday low-energy world, *two* baryons already: the proton and the neutron.

The proton is electrically charged, so why can't it have colour, too? Could we not simply identify the neutron with the quark combination $q_R + q_G + q_B$ and the proton with, say, $q_R + q_G$? Of course, we realize that the colour combination RG is not white: it lacks one blue quark, and so RG behaves like antiblue $\bar{B}$ (because $B + \bar{B}$ = white and $B + (R + G)$ = white, we must have $R + G = \bar{B}$). But why should it bother us that $q_R + q_G$ is antiblue, if we are by no means bothered by the fact that the proton is electropositive?

The most important reason by far is that we have never observed colour charge directly. Remember that observing electric charge directly is extremely easy: just rub a plastic comb on your sleeve and it will pick up pieces of paper by electrostatic attraction. When we rub a comb this way, why does it not become colour-charged and attract mesons? If it did, we could certainly expect dramatic effects because the coupling constant of the colour force turns out to be much bigger than that of electromagnetism. No bare colour is ever seen, not even in particle accelerators that operate at energies equivalent to hundreds of billions of rubbed combs, and we must conclude that the energy difference between the white states and the coloured ones is phenomenally large. So the problem remains: how can we make different particles such as protons and neutrons out of quarks?

Apparently, *there is more than one colour triplet of quarks.* We saw an analogous thing in U(1), where the families are singlets. We had the electron, the proton, the muon, and so forth, with all other electrically charged fermions. Likewise, in SU(3) we have several kinds of quark triplet. Each kind is called, in the corny parlance of particle physics, a `flavour`. We might, upon discovering all these uncalled-for flavours, wonder what's cookin', but Nature cares not about our exasperation, and at least two flavours are needed just to make protons and neutrons. The names of those flavours are `up` (*u*) and `down` (*d*), given in analogy with the isospin degree of freedom (which, in its turn, is a hobbled analogy with spin polarization).

Because the proton is electrically charged and the neutron is not, it is evident that the *u*-quarks and *d*-quarks must have different electric charges. Also, since there are three quarks in a proton, *the electric charge of a quark must be a whole number of one-third proton charges.* Usually, a lot of hoopla is made about this 'one-third', but it is not really that bizarre. The reason for the surprised reaction is, I suppose, that education has hammered the whole-unit charges for the electron and the proton firmly into people's brains, so that the one-third seems odd. From a contemporary point of view, the +1 electrical charge of the proton is a historical accident; it would have been better, with hindsight, to have called it +3, and the quark charges would then have come out at ±1 and ±2. In that formulation, the +3 charge of the proton would be a visible reminder of its SU(3) symmetry.

Possibly you will find this a bit too glib; after all, why then does the electron have the same number of negative charge units, namely −3, in this new form of accounting? Is that also a reminder of a deeper underlying symmetry? We will return to this shortly, and argue that, indeed, there is probably a bigger symmetry that encompasses the electron as well as the members of all other fermion multiplets we have met.

In the meantime, we note that the fact that some baryons are electrically charged whereas others are not forces us to accept the existence of quark flavours that differ from one another in electric charge; moreover, the requirement that a baryon be a white combination of three quarks forces us to conclude that quark electric charges are always whole thirds. Thus, we get a suspicion that 'whiteness' and 'integral electric charge' mean one and the same thing, which is a strong indication of a deeper unity between colour and electric charge.

The way in which the electrical charges are assigned is prescribed by the observation that all three pions couple with both nucleons, and that a meson is a composite of a quark and an antiquark. Thus, the pion triplet must be $(u\bar{d}, u\bar{u}$ or $d\bar{d}, \bar{u}d)$, and each nucleon must contain at least one u and one d. Since a nucleon is a three-quark composite, this means that one nucleon is uud and the other udd. Now let the electric charge of u be $x/3$, and that of d be $y/3$. Because the electric charge of the proton is +1, we require $2x + y = 3$ (assuming that uud is the proton; if we had assumed udd to be the proton, nothing would have changed except the way in which we use the labels up and down, which were arbitrary anyway). Since the charge of the neutron is 0, we must have $x + 2y = 0$, and we conclude that $x = 2$ and $y = -1$, so that the electric charge of the u-quark is $\frac{2}{3}$ and that of the d is $-\frac{1}{3}$. Remembering that the charge of an antiparticle is the opposite of the charge of the corresponding particle, we find that the electric charges of $(u\bar{d}, u\bar{u}$ or $d\bar{d}, \bar{u}d)$ are $(+1, 0, -1)$, which corresponds to (π^+, π^0, π^-), as required.

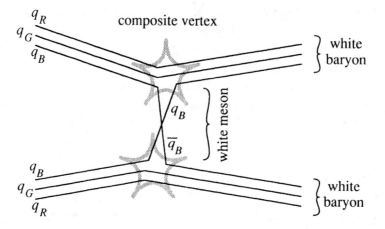

Fig. 11.3 Strong interaction Feynman diagram split up into quarks. When seen on a large scale (in this case, that means roughly the size of an atomic nucleus), the bundle of three space-time trails of the quarks is seen as the trail of a baryon. Similarly, the double space-time trail of the pair of exchanged quarks is seen as another 'single' particle, a pi meson (compare with Fig. 10.10). This exchange of mesons between nucleons corresponds to the Yukawa description of the strong force between particles such as protons and neutrons.

11.3 Mesons and baryons

Are the bound states which form the hadrons similar to those of the atoms, and is there a corresponding periodic table for them? The superposition rules that govern the regularities of the Periodic Table are the same for all particles. Therefore, the quarks in mesons and baryons must be distributed as prescribed by the spherical harmonics. On the other hand, though the quarks are fermions, the Pauli exclusion rule is of no consequence here because all quarks in white hadrons are different. In a meson, they are different because we have a quark and an antiquark; in a baryon, the quarks may have the same flavour but they always have different colours. Therefore, all quarks can be in the ground state and there isn't really a periodic table, the existence of which depends on the fermions being forced into higher energy states (shells) by the Pauli rule. Heavier particles made up of more than three quarks might in principle exist, and would exhibit the kind of classification we see in the Periodic Table. Combinations such as $qq\bar{q}\bar{q}$ and $qqqq\bar{q}$ are white, too, so by the colour rules they could exist. But they are never observed. Calculations of their energy states show that they would fission by the expulsion of $q\bar{q}$ and qqq combinations. An analogy is the alpha-radioactivity or the spontaneous fission of atomic nuclei, in which an entire cluster of nucleons is split off.

There are, of course, excited states with energies above the ground state, just as there are excited states of atoms (we saw that these were responsible for the famous spectral lines of hydrogen). Indeed, particles have been observed in excited states. But a meson or a baryon in such a state will return by spontaneous transition to the ground state in an instant, giving up the excess energy as an emitted quantum. Notice, incidentally, that the distribution of quarks in a white composite is more egalitarian than the distribution of electrical charge in an atom. Because an atomic nucleus is so very much more massive than an electron, an atom is a dictatorship in which the inertia of the nucleus rules: the whole electron distribution is practically centred on the nucleus. But a meson or baryon is more like a co-operative, since the masses of the quarks are closely similar. Because all the quarks in mesons and baryons are in the ground state $J = 0$, they are distributed in a spherically symmetric way. Thus, a proton is a little ball of three quarks.

What hadrons can we expect? That depends on the number of quark flavours. As we saw above, protons and neutrons are made from quarks that have the up and down flavours (u and d). We will see shortly that there are also flavours called charm, strange, top and bottom (c, s, t and b). Evidently, from quarks with these flavours we can concoct a large variety of white composites. A choice of known ones are listed in the accompanying table.

mesons	*composition*	*el.charge*	*spin*	*mass MeV/c²*
π^{+-}	$u\bar{d}$, $\bar{u}d$	1, −1	0	139.57
π^0	$u\bar{u}$, $d\bar{d}$	0	0	134.97
η	$u\bar{u}$, $d\bar{d}$	0	0	548.8
K^{+-}	$u\bar{s}$, $s\bar{u}$	1, −1	0	493.65
K^0	$s\bar{s}$	0	0	497.67
D^{+-}	$u\bar{c}$, $c\bar{u}$	1, −1	0	1869.4
D^0	$c\bar{c}$	0	0	1864.7

baryons	composition	el.charge	spin	mass MeV/c^2
p	uud	1	$\frac{1}{2}$	938.22
n	udd	0	$\frac{1}{2}$	939.57
Λ	uds	0	$\frac{1}{2}$	1115.6
Σ^+	uus	1	$\frac{1}{2}$	1189.4
Σ^0	uds	0	$\frac{1}{2}$	1192.5
Σ^-	dds	-1	$\frac{1}{2}$	1197.4
Ξ^0	uss	0	$\frac{1}{2}$	1314.9
Ξ^-	dss	-1	$\frac{1}{2}$	1321.3
Ω^-	sss	-1	$\frac{3}{2}$	1672.5

By the way, I want to point out that the colour SU(3) symmetry is not the same as the 'eight-fold way' symmetry for the arrangement of baryons. This classification is often trotted out as the prime example of the use of symmetry in particle physics, but I find that a bit misleading. Of course, the mathematical properties of the group SU(3) are the same in both cases, but there the similarity stops. Colour SU(3) has as its fundamental multiplet the coloured quarks (q_R, q_G, q_B), from which all mesons and baryons are made. Eight-fold way SU(3) is based on the coincidence that there are three quarks (u, s and d) that have roughly the same mass, which happens to be fairly low compared with the other quarks. We have already seen, in the discussion of Linnean multiplets, how easy it is to devise different arrangements of given or conjectured ingredients. Likewise, the isospin doublet of proton and neutron did not lead to a correct theory of the strong force. Although it is possible to base a classification system on eightfold-way SU(3), which can even lead to useful predictions, it does *not* produce the right velpons when used as a local symmetry: in this context, SU(3) is not a *gauge* symmetry. But colour SU(3) *is* a gauge symmetry, and the theory based on the exchange of gluons between quarks (called quantum chromodynamics or QCD) is at present one of the mainstays of the standard model of particle physics.

11.4 To kill a pion

When atoms or molecules encounter each other, their outer electron distributions begin to overlap. Thus, there is some probability that two of the electrons end up with the same atomic quantum numbers. If their spin polarizations are the same, then Pauli exclusion ensures that the pair of electrons avoid occupying the same region of space. With opposite polarizations, the paired electrons will be *more* likely to be found in the same region. These effects produce the Van der Waals force, which arises because the shift in the position of the electrons distorts the perfect electric shielding of the atomic nuclei by their attendant electrons. As long as the electron distribution is centred on the nucleus, the electric charges in the atom effectively neutralize each other. But a slight shift leads to an electrical imbalance, so that the atom or molecule acquires an electric dipole field (some molecules, such as water, have a permanent electric dipole, due to the asymmetrical placement of their atoms). These dipole fields can produce attraction or repulsion at distances of a few atomic diameters, but in close encounters the Van der Waals force is always repulsive. Thus, you can put a cup of tea on a table, even though cup, tea and table are almost 100% empty space. The Van der Waals force is a *residual* force, arising when atoms or molecules are slightly bent out of shape.

The strong nuclear force is a residual force, too. When two baryons or mesons approach each other, the encounter causes a temporary shift in the colour charge distribution. This imbalance means that the particles get a glimpse of each other's colour charge, producing a force of the Van der Waals type. The Van der Waals force, being essentially electromagnetic, comes about through the exchange of photons. One might then think that its colour equivalent, the strong force, would be mediated by gluons, but in fact things are a little more complex. The baryons and mesons have such a powerful tendency to remain white that they mostly interact by the exchange of bound combinations of quarks: protons and neutrons are held together in the atomic nucleus by the exchange of quantum beanbags filled with quarks.

Pions, for example, can be exchanged between nucleons, and since their mass is about 270 electron masses, their embezzlement lifetime is brief: only 10^{-23} seconds, corresponding to a range of 10^{-15} metres when moving with the speed of light. That is about the distance between nucleons in an atomic nucleus, and comparable to the nuclear size: about 100 000 times smaller than the region in which the atomic electrons are found. All seems to fit nicely, but for one thing: why don't we observe free pions in our large-scale world? If pions are such nicely bound white particles, where are they?

The answer is that pions are free and long-lived particles when compared with their embezzlement times, but from our gross perspective they die quickly. The lifetime of the π^0 is about 10^{-16} seconds, which means that if one starts with a number of π^0, half of them would have decayed in that

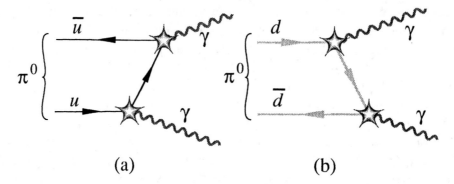

(a)   (b)

Fig. 11.4 Two first-order Feynman diagrams of the decay of a π^0. A pi-zero is made of a quark and an antiquark; at (a), it is an u–$\bar{u}$ pair, at (b) it is a d–$\bar{d}$ pair. Thus, a neutral pion is made up of a matter–antimatter pair; therefore, it can annihilate into two photons.

time. Half of the remainder would be expected to pop in the next 10^{-16} seconds, and so forth. Note that this is a probability prescription, typical quantum behaviour. A more appropriate term for the decay time would be 'life expectancy', but this is not common in the literature. The lifetime of the π^{+-} is vastly longer: about 10^{-8} seconds. Both are much longer than the nuclear timescale of 10^{-23} seconds. If the latter were to correspond to the time of one heartbeat, then a π^0 would live for about four months, and the π^{+-} a whopping thirty million years. For practical purposes the pions are very stable.

It is not too difficult to see how a π^0 could decay (Fig. 11.4), because it is made of a quark and its antiquark. These can be expected to annihilate via the processes $u + \bar{u} \rightarrow \gamma + \gamma$ and $d + \bar{d} \rightarrow \gamma + \gamma$. Two photons must be produced because otherwise momentum is not conserved. However, the π^+ is made of $u\bar{d}$ and the π^- is a bound state of $\bar{u}d$. The flavours of their quarks are different, so how could they possibly decay? Of course it is true that, comparatively speaking, the π^{+-} are very stable. But in the end they do decay, so we must conclude that the u- and d-flavours are, in some sense, like each other. A combination such as $u\bar{d}$ might live forever, except that somehow there must be a small amplitude for processes such as $u \rightarrow d$ or $\bar{d} \rightarrow \bar{u}$. When that happens, we can have the process $\pi^+ = u + \bar{d} \rightarrow d + \bar{d} \rightarrow$ annihilation. We will come back to this in the following chapter.

12

Smashing symmetry

⭐

12.1 Weak decay

Let us return to the problem of pion decay. It is observed that the π^{+-} have a lifetime of 10^{-8} seconds. By subatomic standards that is an immensely long time, but it is finite; and since the π^+ and π^- are made of the quark–antiquark pairs $u\bar{d}$ and $\bar{u}d$, we suspect that the u- and the d-quarks can turn into each other. If they do, we can have reactions of the type $u + \bar{d} \rightarrow d + \bar{d}$, followed by annihilation of matter and antimatter into a blaze of energy. For such a reaction to occur, it is necessary that the distinction between u and d is not absolute: a u-quark is a little like a d. In the quantum world, that does not mean that we are dealing with a particle which is intermediate between these two types of quark. Rather, it means that, for a small percentage of all interactions, the quark behaves entirely like a d; the rest of the time, it behaves entirely like a u (recall the goat–lion chimaera!) If the u–d mixing is only small, the probability that the u-quark will show its d-guise is not large, so that – on the average – we will have to wait a long time before the charged pion can decay: a u-quark is a little like a d, but not very much so, whence the long lifetime of the π^{+-}.

What symmetry can turn one particle into another? If we make the simplest assumption, namely that the two particles do not mix with any others, then they form a doublet. It is then plausible to expect that the symmetry is two-dimensional, just as in the case where we turned a proton into a neutron and vice versa. In this way, we automatically end up with the isospin group SU(2). Then the u- and the d-quark form a doublet under SU(2) symmetry.

When we apply this group as a *local* gauge symmetry, we obtain a triplet of velpons, to be called (W^+, W^0, W^-); it is a triplet because, as you may recall, the number of velpons in an N-dimensional gauge group is $N^2 - 1$, which for $N = 2$ produces three. Then we can have a vertex (Fig. 12.1) of the type $u \rightarrow d + W^+$, and the pion can decay as $\pi^+ \rightarrow W^+ + \gamma$. The corresponding force must be very weak, or else transmutations such as $u \rightarrow d$ would be too frequent and the pion would decay faster than is observed.

Alas, what is observed always looks like $\pi^+ \rightarrow \mu^+ + v_\mu$; that is to say, the reaction products contain at least one particle with mass (in this case, the antimuon μ^+). The v_μ is the mu-neutrino, a massless uncharged particle with spin $\frac{1}{2}$.

Observations soon show that this type of quark conversion plays a role in other forms of particle decay, too. For example, a neutron is a bound state of three quarks, udd; if we have $d \rightarrow u + W^-$, according to the above conjecture, we should expect that the neutron can turn into a proton via $n \rightarrow p + W^-$. This type of reaction can occur spontaneously because the proton p consists of the bound triple uud, and the mass of the neutron n is just a shade larger than that of the proton, so that $n \rightarrow p$ is a reaction in which energy is liberated. However, what is observed (Fig. 12.2) is different, namely $n \rightarrow p + e^- + \bar{v}$: a neutron changes into a proton while emitting an electron and an antineutrino!

A neutrino is a massless uncharged particle with spin $\frac{1}{2}$. It always occurs in conjunction with an electron, just as the mu-neutrino occurs in interactions with muons. This leads to the suspicion that the pair (e^-, v) is a doublet under some two-dimensional symmetry, and that (μ^-, v_μ) is a similar doublet. Naturally, one suspects that these are doublets of SU(2), and that an SU(2) rotation can turn an electron into a neutrino. This behaviour is exactly the same as that which we expected of quark doublets such as (u, d), because of the decay of the charged pions.

Thus, we are led to the possibility that the W^{+-} can have vertices of the type $W^- \rightarrow e^- + \bar{v}$. The resulting neutron transmutation is responsible for the radioactivity called beta decay ('beta particle' is an archaic name for the electron). The process of inverse beta decay is involved (Fig. 12.3) in the occasional capture of an atomic electron by the nucleus: $p + e^- \rightarrow n + v$, which transforms a nuclear proton into a neutron (this process is also called K-capture, because the captured electron comes from the innermost atomic electron shell, the K-shell).

Such processes show that the W-bosons must have mass, or else they could not produce reactions in which all final products have mass. Thus, they cannot be readily identified with the massless SU(2) velpons. We can guess that the mass of the W must be large from the fact that neutron decay is very, very slow: the half-life of a free neutron is 616 seconds. This should be compared with the characteristic nuclear time scale of 10^{-23} seconds

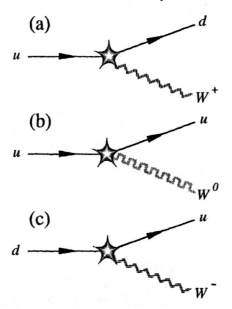

Fig. 12.1 Three weak-interaction SU(2) vertices. The emission of an SU(2)-charged velpon (called W^+, in (a), or W^-, in (c)) leads to the transmutation of a quark from one flavour to another. The emission of a neutral velpon (W^0, in (b)) does not lead to a change in the flavour of the quark.

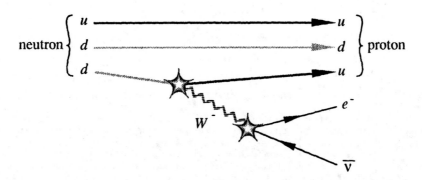

Fig. 12.2 Neutron-to-proton decay resolved into quark paths. The d- and the u-quark flavours mix, that is to say, the neutron contains a particle which behaves d-like most of the time, but u-like some of the time (recall the lion–goat chimaera). Because the weak and the electric charges of u and d differ by one unit, we can have $d \to u$ only when a weak velpon W^- carries away the difference. This boson subsequently decays into an electron and an antineutrino.

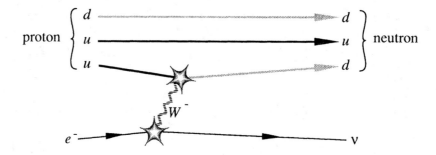

Fig. 12.3 Inverse beta decay resolved into quark paths. This is the opposite of the process shown in Fig. 12.2.

that we encountered when discussing the colour force and its residual, the strong force. If this nuclear time scale corresponded to the duration of one heartbeat, the neutron would have a life expectancy of 10^{26} heartbeats, which is a hundred million times more than the age of the Universe! From this prodigiously long lifetime we conclude that the W-boson finds it terribly difficult to escape from the neutron. This means that the embezzlement time is extremely brief. Accordingly, its mass should be large; a simple estimate† yields $m_W = 60$ proton masses at least. Because of the extremely short range of 10^{-17} metres implied by this, the beta-decay force is called the `weak force`.

12.2 Symmetry breaking

The evidence in favour of treating the weak interactions as a force which is due to local SU(2) symmetry is, first, that we have suitable fermion multiplets for it, namely the doublets (e^-, v), (μ^-, v_μ) and (u, d); second, that local SU(2) produces three W-velpons that appear to describe beta decay

† Let me give you an example how such estimates are made. In order for the W to escape from the proton, it must manage to move at least one proton radius away. The proton radius is given by the Compton length λ_p. But if the W has mass, it cannot travel very far; its range is of the order of its own Compton length λ_W, its embezzlement travel distance. This range does not mean that the W cannot travel farther, but it does mean – true to the uncertainty principle – that it becomes less and less probable that the W will be found more than λ_W from the point where it was temporarily created. In particular, if the Compton length of a particle is λ_C, then its probability density decreases exponentially with distance x, that is to say $P = \exp(-x/\lambda_C)$. To become free from the proton, the W-boson's distance x must be of the order of the 'size' of the proton, i.e. x must be about equal to the Compton length λ_p of the proton. Thus the probability that the W-boson, with Compton length λ_W, is outside the proton is roughly $\exp(-\lambda_p/\lambda_W)$. Now λ is inversely proportional to a particle's mass, so that $\lambda_p/\lambda_W = m_W/m_p$, and $P = \exp(-m_W/m_p)$. As we saw above, only one in every 10^{26} escape attempts is successful, so that $10^{-26} = \exp(-m_W/m_p)$. Because $10^{-26} = \exp-59.9$, this gives for the mass of the W-boson roughly $m_W = 60 m_p$. In other words: because the W is a ball that is at least 60 times smaller than the proton, the probability is at most $\exp-60$ that it will get far enough away from the proton to be effectively separated from it.

properly; third, the circumstantial evidence that the other three known forces all come from local symmetries, namely electromagnetism from U(1), the colour force from SU(3) and gravity from Lorentz symmetry (although the latter is not a quantum theory). Against such an interpretation of the weak interactions we can say that we were once deceived before when we tried to apply isospin SU(2) to the strong force, and that the *W*-bosons are not massless.

The greatest problem is the fact that the velpons created by an exact local gauge twist must be massless.† The reason for this, at which I hinted above, is roughly the following. It was claimed that, because of relativistic effects (Lorentz symmetry), it takes time to cross space, so that we cannot take the concept of a global symmetry seriously. But suppose that all parts of a system were at rest relative to one another. In a world where everything was standing still, all speeds (being zero) would be truly negligible compared with the speed of light. In such a world, the speed of light might as well be infinite: it would not make any difference. The mechanics in this world would be non-relativistic, and we would expect that this would still be the case even if the speeds were not exactly zero, but small compared with the speed of light. In this slow world, which would of course be rather like the world of our everyday intuition, Lorentz symmetry would not be noticeable at all. In that case, a global gauge transformation *would be* possible, so that the dynamical effects of the gauge twist (namely the occurrence of the wrinkle field and its associated force) should *disappear*. This means that it should cost no energy to create a velpon at zero momentum (in a world where everything is standing still, all momenta are zero). The only possible particle that has zero energy at zero momentum is a massless one, and hence the velpon of an exact local symmetry cannot have a mass.

This argument can be connected to the preceding discussions about the relationship between phase changes and forces. You may recall that we produced a kink in the path of largest probability by changing a particle's phase, thereby obtaining Snell's refraction law. Because we identify a deflection with the action of a force, we adopted the Feynman description of the way in which a force acts, namely through its action, a quantity that is proportional to the energy of the force field: (phase change) = $(1/\hbar) \times$ (change of action along the path). We require that the phase change be zero when there is no force, which can only happen

† In writing this chapter, I am recording my version of what, today, is accepted lore. But I must point out that there are dissenters. Perhaps the most distinguished of these is Martinus 'Tini' Veltman, one of the best physics professors I've ever had. I apologize to him for this chapter. For what it's worth, I agree with Veltman that it is distasteful to enlist the help of scalar particles to muck up the behaviour of the vectors.

if the change of the action is zero. However, action depends on energy, and a massive particle *always* has the energy $E = mc^2$ associated with it. Thus, we cannot make it vanish by reducing the magnitude of the local phase rotation: a massive particle always has energy, even if it is standing still. Contrariwise, the momentum and the energy of a massless particle are *both* proportional to $1/\lambda$: thus we can make such particles disappear gracefully by increasing their De Broglie wavelength λ to infinity. We conclude that Lorentz symmetry (which gives us $E = mc^2$) and exact local gauge symmetry (which produces velpons) can only apply simultaneously if the velpons are massless.

This argument would seem to be conclusive, so there appears to be no hope that we can ever construct massive velpons. But there is a fatal flaw in the reasoning. The 'proof' assumes not only that there is an exact gauge group, but also that we must require exact Lorentz symmetry, i.e. it is assumed that there are no preferred directions in space-time. If there were such a preferred direction, then of course space-time would not be invariant under the Lorentz superrotation, just as the pattern on a striped tablecloth spoils its rotational symmetry.

But there can, in fact, be a preferred direction in space-time. You will recall the photon and our discussion of its possible polarizations: because it has no mass, it can only have two polarizations, whereas a massive spin-1 particle can also have transverse spin. In the description of the photon, we chose its direction of motion as a reference for the spin polarization, so that we obtained one spin orientation pointing forward and the other backward. But this is only one possible choice for assigning the degrees of freedom of a photon; other conventions could in principle be used, even though most of them turn out to be mighty awkward in actual calculations. As long as we do not tamper with the degrees of freedom of a particle, we are safe.

The choice of polarization along the direction of motion is a choice that fixes a special direction in space-time. In the quaint language of our subject, this choice is called `fixing a gauge`. Some such choice must be made, or else we cannot perform any definite calculations about observable effects: if we refuse to fix a gauge, we refuse to say what exactly we mean by a photon. To calculate, we need a gauge, just as a builder needs indications such as 'up' and 'North' on a blueprint to position a building correctly. It does not really matter what North is, as long as all blueprints use the same convention. But at the moment we make a choice, we have specified a preferred vector in space-time, and *a deviation from manifest Lorentz symmetry has crept in the back door.*

Gauge fixing is an *apparent* symmetry breaking. When you think about it, you will see that analogous things occur in many everyday cases where calculations are needed on a symmetric system. For example, consider navigation on Earth: when you plot the course of a ship or an aeroplane, you must choose a coordinate system (such as lines of latitude and longitude). However, it is impossible to construct a manifestly spherical coordinate grid on a sphere: there are always funny spots such as the South and North poles, or something similar.

The exact symmetry still exists, but the gauge fixing hides it, as it were. All we do by fixing a gauge for the photon is to say that the photon field embodies a special direction in space-time, and hence that the photon field does not transform in a simple way under Lorentz transformations. Thus the argument that gauge twists always create massless field quanta fails, and velpons need not be massless after all.

That is fine, but is there a natural way in which such a symmetry breaking can be built into our local gauge symmetry? Does it provide our velpon with mass, and how?

12.3 The asymmetry of the vacuum

The answers to these questions are yes and yes. A velpon may acquire mass if we somehow manage to interfere with the perfection of the local gauge symmetry. In the case of atoms, we saw that the symmetry of the electron distribution is broken by interactions between the electron and nuclear spin, and among the electrons themselves. As a consequence, the main energy levels are split. An energy difference corresponds to a mass, via $E = mc^2$. One of the most important consequences of this oft-abused equation lies just below the surface: it says that, because mass is tied to energy, *mass is determined by the environment*, mass depends on how a particle moves; mass is in the context, as it were. Thus, we expect that symmetry breaking – which may entail an energy difference – can lead to the appearance of particle masses. If we want to maintain the symmetry itself exactly, then we have only one choice for introducing deviations: the object that the gauge group acts upon must be asymmetric. This implies that there must be a preferred direction in the vacuum, so that *the vacuum must not be symmetric under some gauge transformations*.

Oof! That is a big lump to swallow. How can we ascribe properties to the vacuum? And in particular, how could it possibly be asymmetric under any group whatever? Well, all we have to do is follow the venerable tradition founded by Descartes. Because the vacuum has physical attributes, if only that of being extended in height–width–depth, it is perfectly sensible to consider it as a physical entity. Thus, let us consider the concept of force yet again. Because the facts of relativity compel us to abandon action-at-a-distance and to accept the reality of fields, the idea of the vacuum as an empty stage on which the Universe's particles play, becomes rather shaky. And because of the facts of quantum behaviour, we have had to assign a very active role to the vacuum, with pair creations and all. Hence, we cannot escape the conclusion that *the vacuum is part of the stuff that the Universe is made of*, and not some sort of pre-existing nothingness without properties.

Once you become accustomed to this very basic idea, you may get to like it. Descartes would have chuckled: *'Je vous l'avais bien dit, mon ami.'* The notion of an active vacuum abolishes the unpleasant problem of the empty vacuum: how could there be such a medium that is not allowed to have any properties, and yet surely has the property that it allows action to take place in it? The vacuum is not some invisible graph paper on to which the Universe is drawn; it is real stuff, and, as we will presently see, it can have properties that are very basic to a description of the forces of Nature.

There is at least one property of the vacuum on which we can probably agree (this may in fact be its defining property): the vacuum is the lowest available energy state. That much is evident because, if it were in a higher energy state, quantum effects would soon nudge it over to make a transition to lower energy, thereby liberating energetic and/or massive quanta. Such a spontaneous transition† occurs in other circumstances as well; we encountered a special case earlier when we discussed the energy levels in atoms. The reason that spontaneous transitions can occur is to be found in the quantum nature of the world. Because any quantum state can be thought of as a chimaera, that is, a superposition of other possible states, it follows that no quantum state is 100% isolated from all other states. If we were to observe a quantum chimaera, we would find that in (say) 15% of all observations it is a lion, in 28% it is a goat and in 57% it is a serpent. Thus, there is always a certain chance, however small, that a given state will be observed. What we call a raven may, for 10^{-20} percent, be a writing desk – the percentage is small, but not zero. Suppose that there is a state among those in a superposition that has a lower energy than some of the others. Then the system will most probably remain stuck in that state, at least until

---

† There is a fascinating possibility that such a spontaneous transition in fact occurred in the first instant of our Universe. This so-called `inflationary transition` might be responsible for many observed properties of the Universe, such as its homogeneity and the rate at which its expansion slows down.

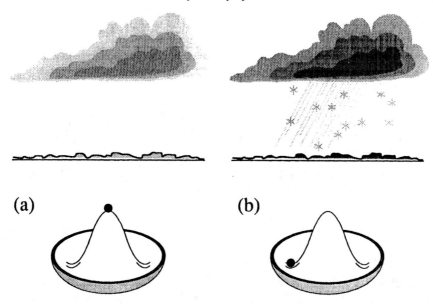

Fig. 12.4 Thundercloud symmetry breaking. (a) No particles, high temperature; (b) particles (snowflakes) appear at low temperature. Below these images are drawn the corresponding solutions for the field. At (a), a symmetric solution with high energy (the ball lies at the top of the potential mountain); at (b), a solution with lower energy, but with broken symmetry (the ball lies somewhere in the valley).

it can garner enough energy to jump back to a higher-energy state. So it is with the vacuum.

Consequently, the vacuum is not necessarily the state that has no particles in it, or even the state with the smallest density of particles. It is perfectly possible that a state with low energy may contain more particles than a state with higher energy. Consider the following analogy: given a cubic kilometre of atmosphere consisting of air and water vapour (Fig. 12.4) and a cubic kilometre with the same matter, but in the form of air and snowflakes (particles!). As it happens, the energy density of the air with snowflakes is *lower* than that of the air with water vapour. The laws of nature that describe air and water have no built-in preference for the presence or absence of snow particles; it's just that at some values of the energy content, snowflakes appear.

Even though the *laws* are symmetric, the *situation* they describe may not be. As soon as the energy is low enough, particles appear spontaneously and the symmetry is `spontaneously broken`.

How could spontaneous symmetry breaking occur in the vacuum? If the symmetries of Nature allow the existence of fields that make the energy of the vacuum *lower* by creating free quanta, then the true vacuum is not the high-energy state without particles, but the asymmetric low-energy state with particles. Such a situation could develop, for example, if the velpons could interact among themselves. The fact that such mutual interactions can make the overall energy lower is well known from other branches of physics. For example, two separate neutrons and two separate protons have a higher energy than a helium nucleus, which consists of two neutrons and two protons bound together. The energetic difference between these two ways of arranging the four nucleons is called binding energy. For example, the mass of two neutrons plus two protons is larger than the mass of a helium nucleus, which consists of those particles bound together by the strong nuclear force. This energy is liberated when, in a nuclear fusion process, helium is formed from hydrogen; it is the energy that makes the Sun and the other stars shine.

The fact that a bound aggregate of particles has a mass that is smaller than the sum of the particles individually does not mean that the individual particles in a bound state have lower masses: the total mass is a property of the aggregate as a whole. Thus, if one were to perform a close-up measurement of the mass of a proton in a helium nucleus the proton mass would still be the same as that of a single proton.

Interaction among velpons is expected in all non-Abelian gauge fields, as was shown before, so invoking that mechanism here is perfectly natural. The details, however, are very complicated, so we will simply assume that there is an interaction which produces the desired effect.

12.4 Velpons can be massive, but...

As we have seen, all exchanges of velpons lead to the appearance of a force, so that the self-interactions of the field quanta generate a kind of overall force that tends to draw them together. It is as if the velpons are woven into an irregular mesh by the interconnections provided by the virtual quanta that mediate the self-interactions. The energy associated with the elasticity of this mesh breaks the symmetry of the vacuum. Like any elastic mesh, the web of interacting velpons carries vibrations. As we know from the De Broglie rules, the frequency of quantum vibrations is directly proportional to their energy, and by the rules of relativity this implies an equivalent mass. Thus, self-interaction becomes responsible for creating massive particles, lowering

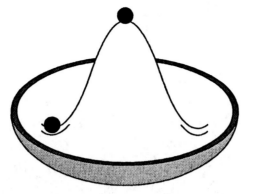

Fig. 12.5 A mechanical analogue of spontaneous symmetry breaking. The solutions of the field equations correspond to the equilibrium positions of a small ball on a sombrero-shaped hill. A symmetric solution has a high energy: the ball lies at the top of the mountain. Another possible solution has lower energy, but with broken symmetry: the ball lies somewhere in the valley.

the energy of the vacuum by the same amount that was invested in these particles.

How can symmetric laws produce an asymmetric product? Simply because symmetric overall conditions need not produce a symmetric equilibrium. For example, consider the orbit of a planet in our Solar System. The laws of motion that describe it are spherically symmetric: in astronomy, it is quite absurd to insist that a certain direction is up, even though most pictures of our planet in space are printed in such a way that Earth is in the lower half of the frame! But a planetary orbit does *not* fill the surface of a sphere. On the contrary, the orbit lies in a plane. One might then suspect that the orbits are at least circularly symmetric in that plane, but not even that is true: they are ellipses!

The existence of an asymmetric solution to a symmetric equation is such a basic fact of life that planetary motions are described in a coordinate system (called ecliptical coordinates) that takes these asymmetries explicitly into account. This is the astronomer's way of 'fixing a gauge'. At the same time, however, the existence of perfectly spherical clusters of stars demonstrates that there are other solutions of the same equations that do show manifest spherical symmetry. Likewise, a house built out of bricks does not have to have the shape of a brick (architects, take note!)

To connect this mechanical example with the quantum problem at hand, consider a related type of motion, namely the path of a tiny ball on a hill that is shaped like a sombrero (Fig. 12.5; see also Fig. 12.4). The shape

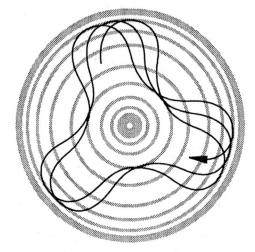

Fig. 12.6 Weaving path of a ball bearing in the bottom of a wine bottle. The path is composed of two simultaneous motions: a constant rolling around the bottom and an oscillating motion across it.

of the hill is cylindrically symmetric, and sure enough there is a symmetric equilibrium state: the ball can be on the exact centre of the central bulge. But this situation is unstable, and a stable, lower-energy solution exists, in which the ball is somewhere in the valley around the central hill. But the ball can be in *any* direction as seen from the top of the hill, so the low-energy solution is patently asymmetric. Likewise, the self-interaction of velpons that are made by an exact gauge symmetry can produce a perfectly symmetric hill of binding energy, and yet generate an asymmetric equilibrium in which free quanta are present rather than absent.

Do the particles produced by spontaneous symmetry breaking have mass? Disconcertingly, the answer is yes and no. Continuing the above analogy, consider the motion of the ball in the neighbourhood of its stable equilibrium position. As was stated above, the symmetry is broken because the ball can be found in any direction as seen from the top of the hill. No position on the circumference of the bottom is more likely than others, by virtue of the fact that the sombrero itself is perfectly symmetric. Thus, the presence of an overall symmetry implies that a particle never feels a restoring force if it is displaced along the circumference of the sombrero. But a small displacement across the valley does generate a restoring force. Therefore, the ball can execute two motions simultaneously: a continuous rolling around the hill and an oscillation across (Fig. 12.6).

If you can snap the bottom off a Bordeaux wine bottle, or somehow obtain a surface similar to the one sketched, you should experiment with the motions of a small ball bearing on this surface. The combination of the two possible modes of motion makes the ball weave about in an undulating pattern as it goes around the central hill. The De Broglie rule $E = 2\pi\hbar f$ and the relativistic $E = mc^2$ say that the frequency f of the oscillation is proportional to the mass of a particle: $m = 2\pi\hbar f/c^2$. Clearly, because we observe that the ball rolls to and fro, there is a frequency and hence there is a mass associated with the spontaneous symmetry breaking.

But there are *two* degrees of freedom, one around the bottom of the hill and one across it, so there are *two* particles. What are their masses? To determine that, we must make a choice: we must fix a coordinate system, a preferred direction, so that we can decide unambiguously which particle is which. This amounts to *fixing a gauge*: it is basically the same procedure that we encountered above in fixing the polarization of the photon, and in precisely the same way symmetry breaking has crept in the back door. The exact symmetry still exists, but the gauge fixing hides it. In the case of the photon, this choice fixes the polarization, singling out one special direction in space-time.

In the case of gauge symmetry, the choice singles out one special direction in the internal space on which the gauge group acts (e.g. electron–neutrino space or isospin space in the case of SU(2)). And, just as with the photon, the choice is basically arbitrary; but it turns out that there is only one clear and uncomplicated way of doing it. When you observe the ball rolling and weaving a complex orbit around the bottom of the wine bottle, you realize that the natural frequencies of the motion are awfully mixed up. In the quantum picture, this means that the states of the two particles are badly messed up and we obtain two quantum chimaeras that constantly turn into one another. There is only one natural and simple gauge choice: to split the motion in an exact radial oscillation across the deepest part of the well and an exact circulating motion around the axis of symmetry of the well (Fig. 12.7).

This choice of gauge fixes two frequencies. One, the frequency f in the radial direction; this, by the rule $m = 2\pi\hbar f/c^2$, corresponds to a particle with a certain finite mass. Two, an unending rotational motion around and around the well; since the speed of the revolving motion is constant, we may say that it takes an infinitely long time for the particle to come to a stop. Because the frequency is the inverse of the time between successive visits to a turning point, the frequency corresponding to the round-and-round motion is *zero*, and that means zero mass. Thus, in this case of spontaneous

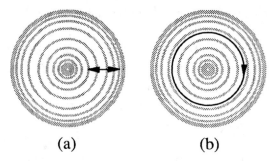

(a)   (b)

Fig. 12.7 Separation of the two frequencies of the motion in the bottom of a wine bottle or the brim of a sombrero: one, radial oscillation with a fixed non-zero frequency (a), the other, motion around and around (b). The latter motion corresponds to zero frequency.

symmetry breaking, one particle has mass and the other one is massless; the massless one is called a Goldstone boson.

Fine, we have a mechanism that can produce massive particles, but at a disastrous price: whenever we invoke spontaneous symmetry breaking, we produce additional massless particles! These are utterly unwanted because they are not observed in Nature. Is spontaneous symmetry breaking a mirage after all? For a moment, it seemed to work so beautifully....

12.5 Cancelling one misery against another

The lament about spontaneous symmetry breaking sounds familiar; we ended on a similar minor chord when we discussed local gauge twists. Usually, miseries in life are strictly additive: you total your car *and* you miss your date. But by a phenomenal stroke of luck, the deficiencies of the local gauge theory can be cancelled by those of spontaneous symmetry breaking. To see the possibility of this, let us list our woes.

Local gauge twists produce velpons that mediate a force. If the symmetry is exact, the gauge twists create massless spin-1 particles. In the case of $SU(N)$, there are $N^2 - 1$ of them. Each massless particle has only two polarizations (namely left- and right-handed helicities), so that the entire velpon field possesses $2N^2 - 2$ degrees of freedom, which must all be used without fail or else we leave cracks in the vacuum. After all, the whole idea of local symmetry is to make up for the fact that the transformation differs from point to point in space-time by introducing a velpon field that compensates for the mismatch produced in the vacuum.

Spontaneous symmetry breaking occurs if the vacuum without quanta has a higher energy than it has when a number of quanta are present. A possible mechanism for this is self-interaction of a field, producing a binding energy between the field quanta. The self-interacting field is not an artificial thing because the velpons generated by a non-Abelian group interact with each other. Since the vacuum is asymmetric in that case, a gauge group acting on it need not produce massless particles. However, spontaneous symmetry breaking produces one massless particle for every massive one.

At this point, we recall that the paramount concern in all local twisting is to make sure that the disturbance created by a local twist is patched properly on to the vacuum at large distances. This patching can be done only if we use all required degrees of freedom in the twist. Thus, the thing that counts is: are all degrees of freedom accounted for? In a sense, the way in which those degrees of freedom are assigned to physical particles is of less concern.

How many degrees of freedom do we need for massless or massive particles? As we saw in the discussion on polarization, the laws of relativity command a massless particle to always move with the speed of light and forbid it to be transversely polarized. Thus, a massless vector boson has two possible polarizations, one with helicity $+1$ (right-handed) and one with -1 (left-handed). A massive particle can never attain the speed of light, and so it can be transversely polarized. Thus, a massive vector boson has two more degrees of freedom, making a total of four: the transverse polarization (helicity zero) contains two possible orientations of the spin vector, at right angles to each other.

Accordingly, we can make a massive velpon out of a massless one if we borrow two degrees of freedom from somewhere. In other words, if we could arrange things in such a way that with every gauge velpon we produce two spin-0 particles, we could reassign the degrees of freedom to combine (massless spin-1) + 2×(spin-0) = (massive spin-1).

But that is precisely what spontaneously broken symmetry gives us! The bothersome Goldstone bosons are just what the doctor ordered: our extra degrees of freedom (i.e. polarizations) are derived from the scalar particles that must occur if the symmetry is spontaneously broken. Remember that there are two possible states of motion in a sombrero potential: one oscillating across (that is to say, a massive particle) and one going around and around (i.e. massless Goldstone particle). We can reassign its two degrees of freedom to a massless SU(2) velpon. This reassignment doesn't make any difference to Nature, as long as we patch up the wrinkles in the vacuum.

In this way, *massless particles disappear altogether*; the symmetry breaking gives mass to the velpons of the local symmetry group and the scalar particles produced by the symmetry breaking provide the extra polarizations that the massive velpons must have. So everybody's happy: local gauge twists with spontaneously broken symmetry can produce vector bosons that have mass, and all degrees of freedom needed to patch up the locally twisted vacuum have been accounted for.

We can connect this reassignment of particle polarizations with our preceding discussion of the relationship between forces and phase shifts. You may recall that the requirement that velpons be massless was based on the equation (change of phase) = $(1/\hbar)\times$(change of action along path); we argued that the action and the phase change must vanish together, or else the equation is false. In that case, we cannot allow massive velpons: their action has a minimum value set by $E = mc^2$. Suppose, however, that we do allow massive velpons, but that we compensate for their bad behaviour by introducing additional scalar particles. Because the action depends on *all* the particles that make up the force fields, it may be possible to arrange things in such a way that the phase equation for the *combined* action of all participating fields is still correct.

Symmetry breaking giveth us a Goldstone particle and the transverse polarization of the velpon taketh it away. This procedure, called[†] the Higgs mechanism, does in fact work. The extra Higgs field is made of scalar bosons, which compensate for the bad behaviour of the massive velpons. They provide the extra degrees of freedom that are needed to produce a massive vector boson triplet.

There is a price to be paid for the Higgs mechanism, namely that the required compensation of degrees of freedom only works in one special gauge, in which one vacuum particle has frequency zero (= the Goldstone boson, going around and around the sombrero). Thus, there must be a special vacuum coordinate system that is specially arranged to make this happen: the symmetry is spontaneously broken.

12.6 The electroweak force

We have now managed to produce a field theory that generates massive velpons which mediate a force. Can it be used to describe the weak interactions? Yes it can, and more: it turns out that it is possible to incorporate the weak force *and* electromagnetism in the same formalism. The germ

---

† Particle physics is a phenomenally competitive pursuit and the names that are attached to mechanisms, effects or theorems are attributions that are disputable. They do not always reflect the real history of the subject. The clever use of scalar particles discussed here was also invented by the Belgian physicists Englert and Brout – but maybe their names are too difficult for Anglophones? For a wealth of historical details, see Pais, *Inward Bound*, and Crease and Mann, *The Second Creation*.

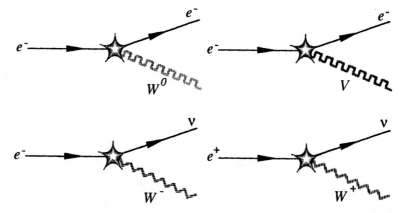

Fig. 12.8 Electroweak interaction vertices with W- and V-bosons. The W-bosons are generated by the SU(2) symmetry, the V-boson by the U(1) symmetry. The unified theory produces four bosons from the product symmetry U(1)⊗SU(2).

of this possibility can be seen in the fact that the basic weak multiplet, the doublet (e^-, v), also contains the basic singlet of electromagnetism, (e^-). This suggests that we use the electron–neutrino doublet in an SU(2) symmetry and use the electron separately as symmetric under U(1).

The hassle with this is that the neutrino is electrically neutral, so that it seems at first that we should exempt it from participating in any combined SU(2)-plus-U(1) scheme. In other words, it does not appear to be possible, at first, to think of the doublet (e^-, v) as the basic multiplet of a product group SU(2)⊗U(1). The way out is the following. Let the SU(2) symmetry produce its vector boson triplet (W^+, W^0, W^-) as before and let the U(1) group produce a 'V-photon' that couples to electrons as well as neutrinos (Fig. 12.8). The W^{+-} carry weak charge as well as electric charge, but the W^0 and V are neutral.

If this were all, then there would be no observable difference between electromagnetism and the weak interaction; after all, if the W^0- and the V-bosons derive from one local symmetry (namely SU(2)⊗U(1)) we should expect them to produce only one type of vertex. In the parlance of particle theory, we say that the W^0 and the V are degenerate with each other; they cannot be distinguished, just as in the case of the electrons in an atomic electron shell. There we saw that each electron state is characterized by three quantum numbers, but only one (the principal quantum number n) appears in the expression for the energy $E \propto 1/n^2$ of a state. The energy levels are then degenerate in the angular momentum quantum numbers l and m:

unless something special intervenes, we cannot distinguish states with the same n but different l or m. This degeneracy is caused by the requirement of exact rotational symmetry in space.

Something has to be done if we want to produce a difference between the weak and the electromagnetic forces. Now we remember the lessons we learned as a quantum sorcerer's apprentice. We can take the W^0 and the V and produce two different superposition chimaeras from them. First, we split V and W^0 into two parts by rotation over some angle: $V \sin \theta_W$ plus $V \cos \theta_W$ and $W^0 \sin \theta_W$ plus $W^0 \cos \theta_W$. Rotating the particles in some abstract space over an angle θ_W should make no observable difference because the sine and cosine mixing functions were chosen precisely so that $\sin^2 + \cos^2 = 1$ and the average of $\sin \times \cos$ equals zero. After this rotation, we swap the sine parts and obtain two new spin-1 bosons, the photon $\gamma = V \cos \theta_W + W^0 \sin \theta_W$ and the neutral vector boson $Z = W^0 \cos \theta_W + V \sin \theta_W$. We choose the superposition in such a way (Fig. 12.9) that the photon couples exclusively to the electron in the (e^-, v) doublet.†

But can we really do this? What is it that is so special about the γ-chimaera? We cannot favour one superposition over another, unless there is something that picks out one particular combination, i.e. one fixed value of the mixing angle θ_W.

At this point, we remember that we also must provide the W^{+-} and the Z with mass, and we know how to do it: by allowing the $SU(2) \otimes U(1)$ symmetry to be spontaneously broken. A broken symmetry means that there is a preferential direction in the vacuum, and it is precisely this that picks out the γ from all other possible combinations. Thus, saying that the symmetry is broken is the same as saying 'the mixing angle θ_W has one specific value'. Only *one* specific combination of the W^0 and the V treats the vacuum as symmetric; thus, that combination has no mass, and is identified with the photon. Consequently, the other combination can *not* see the vacuum as symmetric, and that boson must have a mass.

The photon does *not* acquire a mass. This means that there must be a loose quantum of the Higgs field, a Higgs boson, around somewhere. It has not yet been observed, presumably because it is so massive that it has been impossible to conjure it up in particle accelerators.

In this way, the degeneracy of the W^0 and the V is removed. We saw something very similar in the hydrogen atom. In the ideal case, each atomic energy level (fixed n) is made up of a number of sublevels, each with the *same* energy, but with different angular momenta (l and m). However,

---

† Of course we must extend this coupling to quarks, muons, or whatever else is electrically charged. We must then expect that such particles have partners with which they can make SU(2) doublets.

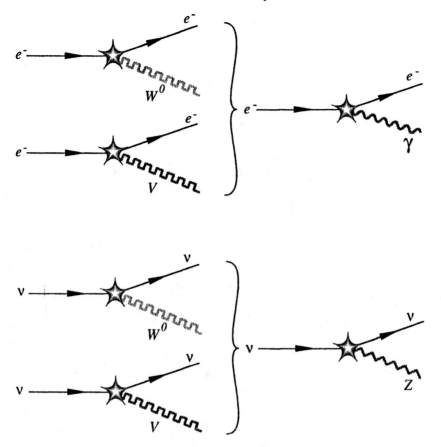

Fig. 12.9 Vertices with W^0 and V combined into vertices of γ and Z. The super-position of the neutral SU(2) boson W^0 with the neutral U(1) boson V is chosen such that one combination produces a massless particle, the photon γ, whereas the other combination, Z, has a finite mass. In the professional literature, one finds the notation Z^0, but because everyone knows that the Z is neutral I've dropped that pesky index.

atomic electrons are not exactly distributed as spherical harmonics because of small perturbations. The three most important of these are: interactions among the electrons; interaction of the spin of an electron with its angular momentum; and interaction between the spin of an electron and the spin of the atomic nucleus. In hydrogen, with its single electron, only the latter two effects occur. These interactions remove the degeneracy of the energy levels: the extra couplings subtly damage the perfect symmetry, so that the different l, m quantum states are no longer equivalent. The symmetry is broken, each principal energy level is split up and we can observe the l and m quantum numbers after all. The interactions of the electron break the rotational symmetry of the atom; the behaviour of the vacuum breaks the symmetry of the weak and electromagnetic forces.

It is marvellous that this construction allows the unification of the electromagnetic and the weak interactions into one `electroweak force`. But it is a bit disconcerting that the electroweak symmetry group is a product group SU(2)⊗U(1), and not a single all-encompassing one, such as SU(3). Accordingly, we cannot really say that electromagnetism and the weak force are the same in the sense that electricity and magnetism are. The fact that they are only quasi-the-same is responsible for the somewhat peculiar and not very straightforward way in which the velpons are constructed. The multiplication of the U(1) group with SU(2) means that the velpons produced by them must be mixed states, obtained by the superposition of V and W^0, producing quantum mechanical chimaeras with a bit of the properties of both. Since the V-photon is charge-neutral, it can only form a state that can be identified with an observed particle by combining with the W^0, which is likewise neutral. Because the number of degrees of freedom must be rigorously accounted for, the two vector bosons V and W^0 can only be combined to give two mixed states: one is the electromagnetic photon γ, the other is a neutral particle called Z. The superposition of V and W^0 is done in such a way that the photon couples only with electrons and not with neutrinos. Why the vacuum of our Universe behaves like this, only Nature knows – so far.

Thus we end up with four electroweak velpons: the massive vector boson triplet (W^+, Z, W^-) and the massless photon singlet γ. (By the way, in the professional literature the Z is called Z^0; but there is no charged Z anyway, and I refuse to type in all those superfluous noughts). But there is more: because we have invoked spontaneous symmetry breaking, we have used four Higgs particles to give mass to the W and Z. But since the photon remains massless, there must be an unused Higgs particle and an unused Goldstone boson around somewhere! Because the mass of the W and Z is of the order of a hundred proton masses, we do not expect that the mass of our mystery guest is less, and possibly it is much more; so far it has escaped detection, but the W- and Z-particles have now been found, in the expected mass range (81 and 94 proton masses, respectively).

One of the peculiarities of the SU(2)⊗U(1) theory of the electroweak field is the existence of the electrically neutral Z-particle. Thus, the elastic scattering of an electron and an antineutrino proceeds not only via the electrically charged interactions but also via the exchange of a Z: These `neutral current` processes (Fig. 12.10) have been observed and formed the first tangible evidence for the correctness of the SU(2)⊗U(1) construction.

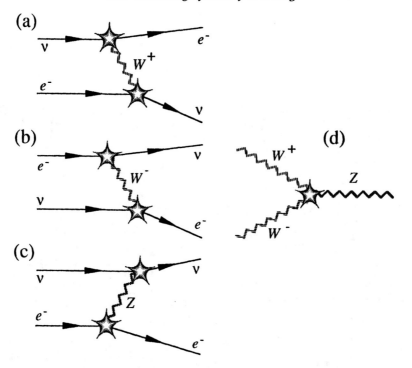

Fig. 12.10 (a,b) Electrons can scatter off neutrinos by the exchange of a charged *W*-boson. This is called a charged-current process. (b) By the exchange of a neutral *Z*-boson, scattering can occur in a neutral-current process. In (d) we see that, because of the non-Abelian character of SU(2), the velpons carry weak charge and can interact by $W^+ + W^- \rightarrow Z$.

12.7 Observing symmetry breaking

Is there any directly observable consequence of the fact that the SU(2) symmetry is broken? There ought to be if Noether's theorem, according to which every symmetry produces a corresponding conservation law, can be extended to imply that a broken symmetry produces a broken conservation law. Let us see if the weak interaction shows signs that the relevant conservation law (the conservation of weak charge) is imperfectly obeyed. Immediately, we are faced with a problem: if weak charge is not conserved, where does it go? There may be many ways in which charge can magically disappear, but only one has been identified so far. To see how this works, consider the analogy of the tablecloth world again.

Associated with any local symmetry is a conserved quantity. In the cases mentioned up to now, these are the electric, colour and weak charge, and the gravitational mass. The twisting of our tablecloth is likewise associated

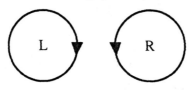

Fig. 12.11 Negative (R) and positive (L) circulation loops serve as an analogy for a certain kind of charge. A unit of 'positive mock charge' is associated with the R-loop, and a negative one with the L-loop.

with the conservation of 'mock charge'. Because mock charge originates in a two-dimensional rotation (Fig. 12.11), we symbolize positive and negative charges by means of circles with arrows. Every creation of a mock charged pair produces one of each of these charges. The symmetry is unbroken, and we have mock charge conservation.

But now imagine that a higher-dimensional being were to pick up one charge, flip it over in three-dimensional space and put it back (Fig. 12.12). Then we would have, for example, a total mock charge of −2 units where previously we had zero. By allowing a three-dimensional flip, we deviated from strict two-dimensional rotation and have broken our symmetry, thereby whisking away two units of charge! Unless our system has some memory of the space flip, the charges have simply disappeared into the vacuum, leaving no trace.

Do we have a physical quantity that behaves according to the model sketched above? Yes: it is the helicity or handedness of a particle. When you see a particle with right-handed polarization coming straight at you, its spin rotation looks like the positively mock charged particle above and left-handed spin is like the negative mock charge. This allows us the following way of undermining weak charge conservation: we stipulate that *only particles with one particular handedness carry weak charge*; others have weak charge zero.

Note that this is an assumption. The association of weak charge with helicity certainly means that weak charge cannot be conserved because the helicity of a massive particle is not conserved. It can be changed by bringing the particle to a standstill and reversing its direction while keeping the spin unchanged. However, we know of no *necessary* association between weak charge and helicity. (Here's a chance for a Nobel prize!). As it happens, it is observed that only particles with left-handed (L) helicity and antiparticles with R-helicity carry weak charge. All others have weak charge zero, and therefore do not feel the weak force.

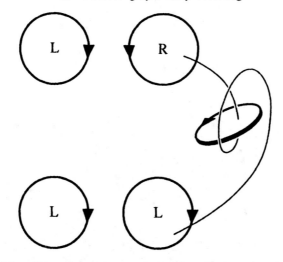

Fig. 12.12 Negative (R) and positive (L) circulation loops can be interchanged if we flip them in an extra dimension. By flipping an R-loop, it is replaced by an L-loop; if we associate a unit of 'mock charge' with the R, we see that this flipping makes two units of mock charge vanish.

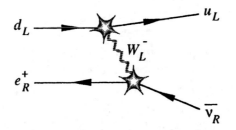

Fig. 12.13 Feynman diagram of a d-quark scattering off a positron by exchange of a W^--boson.

Now we can see straight away how the non-conservation of weak charge would present itself to us. According to the above rules, particles such as an L d-quark or an R positron carry weak charge, so that we can have a process such as that shown in Fig. 12.13, wherein the weak interaction changes the d-quark into a u-quark. (Notice that we start with one L and one R, but end up with two L-particles, for a difference of two units of helicity). However, the process $d_L + e_L^+ \rightarrow u_L + \nu_L$ cannot occur because the L-positron has zero weak charge. If, by way of a Lorentz rearrangement of

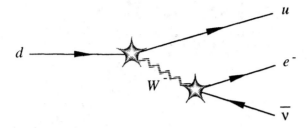

Fig. 12.14 The same diagram as Fig. 12.13, but with the incoming positron replaced by an outgoing electron.

time order, we change the incoming positron into an outgoing electron, we obtain the situation drawn in Fig. 12.14.

As we have seen before, a neutron consists of two *d*-quarks and one *u*-quark, whereas a proton is made up of two *u* and one *d*. Thus, the above reaction can turn a neutron into a proton, an electron and a neutrino. This process, in atomic nuclei, is responsible for the kind of electron-emitting radioactivity called beta decay. If weak charge is not conserved, we expect all beta-decay electrons to have left-handed helicity. Laboratory observations show that this is indeed the case: the electroweak symmetry is broken!

Only particles with mass can change their helicity. The velpons that carry weak charge must have mass, otherwise their helicity would be strictly conserved and SU(2) would not be broken. Thus we see that one phenomenon appears in three guises. First, in abstract mathematical form: the symmetry is broken. Second, in physical but hidden form: the weak vector bosons have mass and therefore can change their helicity. Third, in visible form: Nature is not symmetric with respect to left and right, in that beta-decay electrons have L-helicity. But why weak charge is associated with helicity is not yet known.

This summary also allows us to understand why, at high energies, the effects of symmetry breaking are not readily apparent. When the energy of a process is far above the rest mass energy mc^2 of the weak velpons (say above 500 GeV or so), the energy of the motion of these particles is so overwhelming compared with their rest mass that they might as well have no rest mass at all. In other words, at higher and higher energy a massive particle behaves more and more like a massless one. The helicity of a massless particle cannot be changed; it is very difficult to change the helicity of a very high-energy particle. Changing the helicity is effectively the same as stopping the particle and, keeping the spin unchanged, accelerating it in the opposite direction. If the helicity is conserved, so is the weak charge; accordingly, at high energies the effects of symmetry breaking effectively disappear, and full SU(2) invariance is observed.

Finally, one might wonder what happens to a particle with intermediate helicity. Here we should recall one of the principles of quantum behaviour: namely, that an intermediate quantum state is a superposition of pure states. Thus, a particle with a polarization somewhere between L and R must be thought of as a quantum chimaera, which consists of an amplitude l of an L-particle superposed on an amplitude r of an R-particle. When we observe this object, we see the particle with L-helicity in l^2 percent of all detections and the same particle with R-helicity in r^2 percent. The intermediate nature of the quantum state shows up in the probability with which L-helicity or R-helicity emerges from an observation, not in the helicity itself being intermediate between L and R.

12.8 Weakness and charm

So far, we have talked about the electroweak doublet (e^-, v), but I have already mentioned that the fact that many other particles feel the electromagnetic coupling means that they should be subject to the weak force as well (if electromagnetism and weak interaction can indeed be unified). Quarks carry electric charge, so we might expect that there are SU(2) quark doublets. As an amusing aside, I will present one specific case of weak meson decay that led to the discovery of another quark flavour. Not only does this show the weak interaction at work, but it is also an interesting illustration of the fact that a process must always be seen as the sum of all its Feynman diagrams, and not just one specific diagram.

The decay of the charged pions, for example $\pi^+ = u + \bar{d}$, led us to suspect that there must be a small amplitude for processes of the type $u \rightarrow d$ (Fig. 12.15). If such processes can occur, we may have $u + \bar{d} \rightarrow d + \bar{d}$, after which the antiparticles annihilate each other. Where have we seen this turncoat behaviour before? In all discussions of multidimensional groups, of course. If the u- and d-quarks form a doublet, then we expect them to belong to a two-dimensional group that turns u into d. This group is our old friend SU(2) which, used as a local symmetry, produces $2^2 - 1 = 3$ velpons, the W- and Z-bosons. This symmetry then prescribes vertices such as $u \rightarrow d + W^+$. Such processes would allow pions to decay as $\pi^+ \rightarrow \gamma + W^+$ (Fig. 12.16).

Splendid, but for the fact that there are no W-bosons among the decay products of pions! Instead, we observe an antimuon and a muon neutrino: $\pi^+ \rightarrow \mu^+ + v_\mu$ (Fig. 12.17). The simplest interpretation is that the W itself

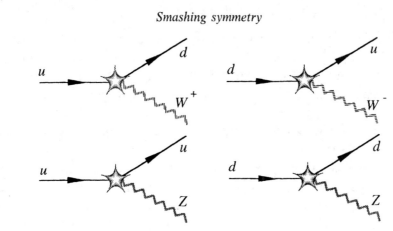

Fig. 12.15 Four possible vertices for the weak interactions of the *u*- and *d*-quarks.

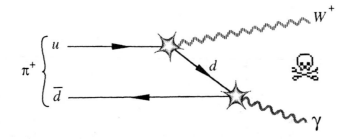

Fig. 12.16 The decay reaction $\pi^+ \rightarrow \gamma + W^+$ of the positive pion is not observed.

splits up as $W^+ \rightarrow \mu^+ + v_\mu$. But why? The *W*-particles are expected to be spin-1 quanta, like the photon. How could they possibly decay into products that contain a massive particle? Clearly, the behaviour of the pion points to an underlying complication. In order to understand the death of the charged pions, we have to anticipate the following chapter a little.

We must never forget that the behaviour of a force is due to the *collective effect of all possible diagrams*, not due to one Feynman diagram only. This collective effect can be constructive, but there are cases where destructive interference between Feynman alternatives causes interesting behaviour.

Quarks can change their flavour at a weak interaction vertex; in other words, quarks carry weak charge. In particular, the electric charges of the quarks *u*, *d* and *s* are $\frac{2}{3}$, $-\frac{1}{3}$ and $-\frac{1}{3}$ respectively; their weak charges are $\frac{1}{2}$, $-\frac{1}{2}$ and $-\frac{1}{2}$. The electric charges of W^+ and W^- are $+1$ and -1, and their weak charges are $+1$ and -1 also. Thus, the electroweak charge of

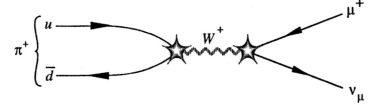

Fig. 12.17 The process $u + \bar{d} \rightarrow \mu^+ + \nu_\mu$ is the reaction which actually occurs in the decay of the positive pion.

the d-quark is exactly the same as that of the s-quark; therefore, in some sense these particles are the same in electroweak processes. Instead of having either a d-quark or an s-quark, with nothing in between, we must think of these particles as superpositions of each other, some sort of quantum chimaera that can sometimes behave as a down quark and sometimes as a strange quark. A transition $d \rightarrow s$, or the other way around, can take place spontaneously by a change in the d-s superposition mix; a change of this type is called a `Cabibbo rotation`.

We can invoke a Cabibbo rotation to turn a u-quark into a d-quark. Then we can have $d + \bar{d} \rightarrow W^+$ in the decay of the π^+ discussed above, and thereby obtain a small chance that π^{+-} will decay; the probability is a hundred million times less than that of the π^0 annihilation, which can occur straight away without the need for turncoat quarks.

This type of chimaera-stirring sometimes produces fascinating results. The collective cancellation of a batch of Feynman diagrams can occasionally be used to infer the existence of unknown quantum numbers or particles! If a particle process that ought to occur is not seen, or is so severely suppressed that it is much more rare than expected, we can search for diagrams that interfere destructively with the process in question.

A case in point is the decay of the so-called long-living neutral K-meson, called K_L^0. This meson was known to consist of a down quark and a strange antiquark; it was expected to decay into two muons via the process shown in Fig. 12.18, involving the neutral Z-boson ('neutral current' process). What is going on here is an example of a vacuum process that produces a d-to-s Cabibbo rotation. This process is seen to occur in the diagram of K_L^0 decay; the s-quark then annihilates with the $\bar{s}$-antiquark, forming a neutral Z-boson.

However, the K_L^0-meson is never observed to decay into a muon–antimuon pair; the accuracy of the relevant high-energy experiments approaches one in a billion. A possible way out is to presume that the above decay diagram

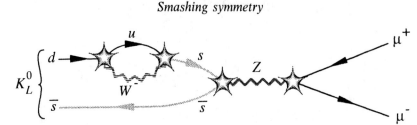

Fig. 12.18 *Z*-decay of the neutral kaon. A K^0-meson consists of a *d*- and an *s̄*-quark. Because the *d* can change its flavour into *u* by emitting a W^-, and because the *u* in its turn can change into an *s*, there is a certain probability that the K^0 decays because *s* and *s̄* can annihilate each other into a *Z*-boson. However, this process is not observed.

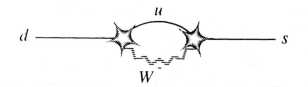

Fig. 12.19 $d - s$ Cabibbo rotation which causes the down quark to change into a strange quark. This process is part of the decay path of the neutral kaon (Fig. 12.18).

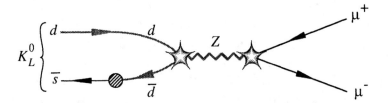

Fig. 12.20 Counterdiagram to Fig. 12.18. In this case, there is a Cabibbo rotation in the path of the *s̄*-quark, shown as a hatched circle. This diagram exactly cancels the one where the *d*-quark turns into an *s*.

is cancelled exactly through quantum interference with the process shown in Fig. 12.20. Here, the Cabibbo rotation changes the *s̄*-antiquark into a *d̄*-antiquark, which then annihilates against the *d*.

Now in order to cancel exactly against the first K_L^0-decay diagram, the Cabibbo rotation must involve a quark with the *same* electroweak charge

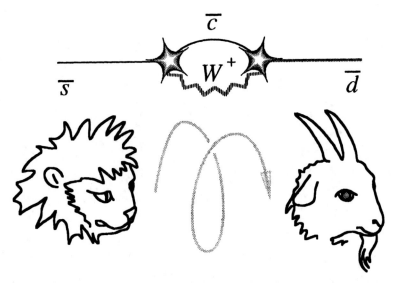

Fig. 12.21 $\bar{s}-\bar{d}$ Cabibbo rotation. If we are dealing with a quantum chimaera, there is always a certain probability that the beast behaves like a lion at one time and as a goat at another.

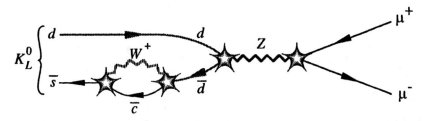

Fig. 12.22 Full kaon-decay counterdiagram, in which the rotation shown as a hatched circle in Fig. 12.20 is expanded into a loop involving a W^+-boson.

and mass as the u-quark. At the time the above cancellation mechanism was cooked up, such a quark was not known to exist, so it had to be invented; it was called the charm quark c. Using this new particle, we can change $\bar{s}$ into $\bar{d}$ via the vacuum process shown in Fig. 12.21.

In this way, the behaviour of the K_L^0 is governed not only by the Feynman diagram in Fig. 12.18, but also by the diagram in Fig. 12.22. Because u, d, s and c all belong to the same family, this diagram suppresses the preceding one exactly: effectively, one diagram is the same as the other, except that the fermions at the vertex are interchanged. Thus, by Pauli exclusion, the amplitudes of the two processes are equal and opposite, so that they cancel. In this reaction, the c-quark is a virtual particle, but after this prediction its existence was confirmed by the discovery of a meson consisting of the pair $c\bar{c}$.

13

How much is infinity minus infinity?

13.1 Dressing up

Now that I have introduced the symmetries that rule the vacuum, it is time to compare the behaviour of the various forces which these symmetries generate. When we do this, we must realize that such a comparison means that we will have to consider the behaviour of the forces at very small distances as well as at very large ones. Because of the need to include all length scales, we are obliged to treat quantum interactions in a global way. We have seen in the discussion of Feynman diagrams that splitting a relativistic quantum phenomenon into pieces that are too small makes the bits individually unmanageable. Globally, however, they can often be handled.

We must think of the behaviour of a force not as due to one Feynman diagram, but as due to the *collective effect of all possible diagrams* in which gauge twists of a certain symmetry are involved. The individual diagrams are indistinguishable: only the in- and outgoing states of the particles are fixed. What happens in between is *indeterminate*. Thus, we get interference between the various alternatives. This interference, as always, can be destructive or constructive. As we will see presently, this has a profound influence on the way in which a force depends on the distance over which it acts.

Let us first consider the propagation of a single electron. Because there is no universal time order of events, a sequence such as the one shown in the top half of Fig. 13.1 (scan with your Feynman diagram scanner!), which appears to be a series of multi-particle events, may to another observer look like a single particle, as is apparent in the bottom half of the diagram. Therefore, if you allow interactions between electrons and positrons, Lorentz symmetry compels you to allow interactions of the electron with itself. Now it just so happens that the amplitude of the above process is *infinite*. This unsettling

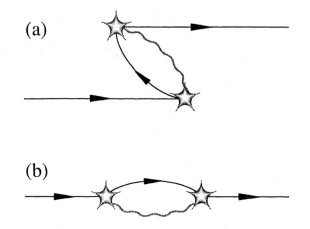

Fig. 13.1 Two ways of looking at the vertices in the propagation of an electron. What seems to be a sequence involving many particles (a) may, to another observer, seem much simpler (b). The sequence at (a) appears to involve particles that sprout from the vacuum, interact with the electron and vanish again into the vacuum. From the viewpoint of a different observer the same diagram may look as if the electron temporarily splits off a photon.

bit of news can be intuitively understood by bringing the two vertices closer and closer together. Then the propagation time Δt between them becomes smaller, so that (because of the uncertainty relation $\Delta E \, \Delta t > \hbar$) the energy of the virtual photon can increase without bounds. Thus, the interaction of an electron with itself gives to that particle a self-energy that appears to be infinite!

This would seem to spell the doom of our theories, but we are saved by adopting an overall view. The global approach allows us to group the Feynman diagrams together in even bigger classes, so that it is assured that collectively they behave properly, even though individually they are very ill-behaved. This is exactly the same idea as grouping vertices together in one diagram; it is a case of 'I'm not OK, you're not OK, but together we're fine.' That is the way of interference.

The grouping together is called renormalization and it works. For example, the infinite self-energy of the electron can be brought to heel by summing the series of diagrams in Fig. 13.2. In rough outline, this works as follows. Suppose that we are trying to get an electron from vertex V to vertex F. In the classical case, the only such process is simply a straight line from V to F (the first path in the diagram). This is what is called a naked electron. But in the quantum case, all we know is that this path is the

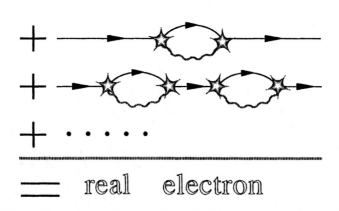

Fig. 13.2 Series with an increasing number of electron–photon loops. The propagation of a real electron is made up of the sum of all possible loop diagrams that have given, fixed, in- and out-states. What we gross beings in our big laboratories perceive is the naked electron plus all its attendant vacuum loops. By ascribing all these loops to that one propagating electron, we obtain a finite result for all observable quantities.

most probable; if Δt is the time of flight from V to F, then deviations in the energy of the process are possible, provided that the product $\Delta E \, \Delta t$ is constant, so that $\Delta t = \hbar/\Delta E$.

The longer a certain deficit can be maintained, the more probable it is. Thus, it is plausible that the amplitude of going from V to F with an energy deficit ΔE is proportional to Δt, namely $P =$ (amplitude) = constant $\times \Delta t = A/\Delta E$, where A is some constant that depends on the details of what happens at a vertex. Thus, the diagram with one photon loop from V to F is associated with the amplitude P. But then the diagram with two such loops in succession has amplitude P^2, because – as we know from everyday life – the compound probability of unrelated events is the product of the individual probabilities. Taking the sum over the whole sequence, we obtain (overall amplitude) $= (P + P^2 + P^3 + \cdots)$. Using the expression for P, we can then write that the overall amplitude is proportional to $(A/\Delta E) + (A/\Delta E)^2 + (A/\Delta E)^3 + \cdots$. Using a little algebra it can be shown that this series adds up to $A/(\Delta E - A)$. You can check this summation by adding, on your calculator, $1/2 + 1/4 + 1/8 + \cdots$ until you have reached a sum $1/(2-1) = 1$ (or until you're tired, whichever happens first).

The above argument shows that the amplitude between the vertices is proportional to $1/(\Delta E - A)$. This opens the following interesting prospect:

by summing all the sequential-loop diagrams, we obtain an electron enclosed in a photon cloud: a `dressed electron`. We can keep the amplitude for the propagation of a dressed electron *finite*, provided that we can prescribe ΔE !

Now that sounds odd at first. After all, because $E = mc^2$, prescribing the energy is the same as prescribing the mass. Are we free, then, to do this? The answer is yes, because all we wish to adjust is the mass of the naked electron, that is, the `bare mass` between V and F, and the quantum behaviour says that we cannot observe this region anyway (without upsetting the propagation of the electron). All that is required is that the mass of the dressed electron, outside the region $V \rightarrow F$, be equal to the observed mass of the electron. This trick is called `mass renormalization` and it has been proved that it solves all self-energy problems of the electromagnetic interaction. The remarkable thing is that *one* adjustment of the mass works for *all* self-energy diagrams, not just the ones shown above; but I will not prove that here.

The propagation of a real electron is the sum of all possible loop diagrams that connect given in- and out-states. What we gross beings in our big laboratories see is the naked electron plus all its attendant vacuum loops. By ascribing all these loops to that one propagating electron, we obtain a finite result for all observable quantities.

A point worth noting in the above is this: the propagation of a particle determines the energy shift ΔE and hence its mass. Thus, a particle can acquire mass, as it were, by the way it propagates. *A prescription for the possible interactions of a particle can determine its mass*; we saw this effect in action when we discussed the weak force. The key element here is that a mass is not what our prejudices say it is: we must refrain from making statements about internal processes that we cannot tell apart, because they are indistinguishable alternatives. Instead, the mass is a number in the equation which relates a phase change to a change in the action density. This phase change, in turn, determines the shape of the path that is the most probable. The most probable is determined by a sum over all possible Feynman paths, with action densities that include loops, excursions to Sirius and what not, so it is entirely reasonable that the interactions which govern these loops should have a say in what the particle's mass is.

The most important aspect of the famous relativistic equivalence of energy and mass is that the mass of a particle is determined by the environment in which it appears. All interactions in which it participates help to determine its mass. For example, in the case discussed above, the fact that a propagating electron can temporarily split off a photon means that it gets a mass that

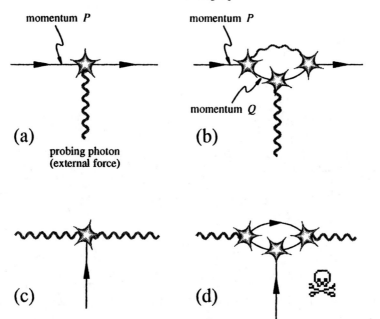

Fig. 13.3 The mass of a particle depends on the circumstances of its propagation. This is the gist of the relativistic equivalence of mass and energy. When an external photon hits the electron when it has split off a virtual photon (b), the probing photon acts on the electron while it has an energy-momentum different from the original one at (a). The response of the electron is then different: it acquires a different mass. This mechanism does not change the mass of a propagating photon (c) because a diagram in which a probing electron interacts with a temporarily created electron–positron pair is forbidden (d).

is different from the mass of a naked electron. We can see this as follows. Let an external force interact with the electron by hitting the electron with a photon coming from outside: see Fig. 13.3. But if it just so happens that the electron has split off a virtual photon before it was hit, the probing photon acts on the electron while it temporarily has an energy-momentum that is *different* from the original one because it has shed a 'bypass photon'. Obviously, the response of the electron is then different. As we have seen from the beginning, mass is determined by the response to an external force (remember the ball and the railway engine). Therefore, the 'bypass photon' produces a mass correction. The deep implication of $E = mc^2$ is: mass is a matter of the environment!

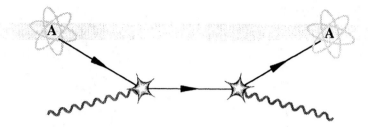

Fig. 13.4 Propagation of light through matter. The photon interacts with electrons in atoms in its vicinity. The photon phase is changed, which means that there is an effective force acting on the photon. Thus, the matter slows the photon down; it acquires mass through the interaction with its environment.

13.2 Through a glass, slowly

In order to see how the propagation of a particle can determine its mass, it is instructive to look at the motion of a photon through matter (say glass). You have probably heard that light travels with the speed c in vacuum only, and that it travels more slowly in a medium. As a beginning student, I was much puzzled by this: on the one hand, I had heard that massless things such as photons always move with the speed of light. On the other, here were photons that travelled *slower*! Did that mean that they have mass?

Well, yes, in a sense. Consider what happens when a photon traverses a piece of matter (Fig. 13.4). How does it know that there's something out there? By interacting with the resident charged particles. The easiest ones to interact with are the electrons, because of their small mass. Every time such an interaction occurs, the phase of the photon changes. Thus, the 'muddy wheel' which traces the photon path has to turn a little more, on account of all these extra side-trips. As we have seen, a change of phase is tantamount to the action of a force. Because of the collective effect of the electrons in the glass, the inertia of the photon changes: it acquires mass, i.e. it propagates more slowly than in vacuum.

We can now understand in a little more detail how the W- and Z-bosons of the electroweak interaction acquire their mass, and in particular what difference it makes that the symmetry is broken. As in the case of the photon, all diagrams of the closed-loop type leave the mass of the velpon unchanged. However, if the symmetry is broken, the vacuum can actively participate in the propagation by means of open processes such as the one shown in Fig. 13.5. What happens here is that the propagating W-boson intercepts a particle H, with mass m; at the first vertex, the symmetry twists

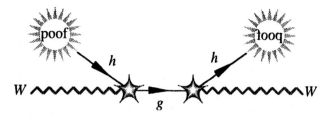

Fig. 13.5 A *W*-boson propagates through the vacuum. Because of spontaneous symmetry breaking, a vacuum with particles is more likely than one without. Higgs particles are created by the vacuum, connect to the *W* at a vertex and an instant later are reabsorbed into the vacuum. This process of 'pushing against the vacuum' changes the propagation of the *W*, giving it mass.

the vacuum, so that *W* absorbs *H* to form a massless boson *G*. At the second vertex, the vacuum untwists, reproducing *H* and *W*. The novelty is that the particle *H* is produced and reabsorbed *by the vacuum*, as symbolized by the circles; this is possible because the energy of the vacuum with the *H*-particle is *smaller* than without it, i.e. symmetry is spontaneously broken. The net result is that the *W*-boson moves as if it had the inertia of the *H*-particle: the *W*-boson has acquired mass by 'pushing against the vacuum', in a manner analogous to what happens to a photon in glass.

One may be worried by this because it seems as if momentum is not conserved: it could only be conserved if the interaction with the vacuum were recoilless. But that is precisely what it is! The particle *H* is produced by the entire Universe, from here to the quasars; the *H* is a collective mode of the vacuum, and since the inertia of the Universe is as good as infinite the process in the diagram is possible with recoilless creation and destruction of *H*. Incidentally, this contact with the vacuum also serves to make it a repository for the weak charge that is spirited away in weak processes which do not conserve helicity. The vacuum is, indeed, the embezzler's paradise.

13.3 Vacuum polarization

Having considered the self-energy of the electron, let us take a look at the propagation of the photon. The naked photon just moves straight ahead. But because the number of particles can change due to spontaneous pair creation in the vacuum, the photon can also dissociate into a particle–antiparticle pair. This, too, has an amplitude that can increase without bounds. The appearance of pairs out of the vacuum, in the path of the photon, is called `vacuum polarization`. Confident because of the success of grouping

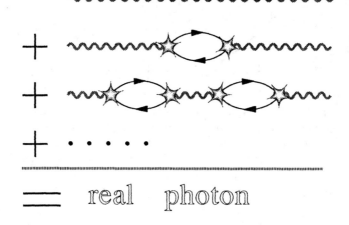

Fig. 13.6 Vacuum polarization series with an increasing number of electron–positron loops. The propagation of a real photon is the sum of all possible loop diagrams that have given, fixed, in- and out-states. What we see on a large scale is the naked photon plus all its attendant vacuum loops. By ascribing all these loops to that one propagating photon, we obtain a finite result for all observable quantities.

Feynman diagrams together, we consider what happens if we add up the series obtained by allowing an increasing number of electron–positron loops in the photon path. These are indistinguishable because all we know from the outside is that the photon goes in and comes out with the same momentum.

This procedure is carried out exactly as before, except for the fact that we have already fixed the mass of the electron (and therefore that of the positron). The probability for each loop is equal to some constant B, which depends on the details of what happens at a vertex. The overall result of the above batch of loop diagrams is then equal to a constant C times the series $1 + B + B^2 + \cdots$, which adds up to $C/(1 - B)$. The number C depends on the strength of the coupling at a vertex. This strength is the electric charge e; since each electron–positron loop has two vertices, it is found that C is proportional to e^2, so that the amplitude for the propagation of a photon is proportional to $e^2/(1 - B)$. This factor does not contain a term of the form ΔE, so that we cannot save things by means of mass adjustment: the mass of the photon remains zero.

However, because the amplitude factor contains e and B, we can adjust the charge of the electron between two vertices in such a way that the vacuum polarization remains finite. This is called `charge renormalization`. We are free to do this as before: all we wish to adjust is the `bare charge` (between vertices), which is a region we cannot observe anyway without devastating the propagation of the photon. All that is required is that the dressed charge that appears when measured from a large distance be equal to the observed electron charge. Again, I state without proof the remarkable fact that one adjustment of the charge works for all vacuum polarization diagrams, not just those shown here. By means of the adjustment of the (unobservable) bare mass and bare electric charge, the bad behaviour of individual Feynman diagrams disappears when they are considered collectively in the sum over all diagrams. The theory of the electromagnetic force is *renormalizable*.

By the way, this mechanism does not change the mass of a propagating photon because a diagram in which a probing electron interacts with a temporarily created electron–positron pair is forbidden (Fig. 13.3).

13.4 Global behaviour of electromagnetism

The active role of the vacuum in the shaping of force behaviour is dramatically apparent in the above. The combination of relativity and quantization produces a microworld in which the number of particles is no longer constant. In the banking system of this embezzler's paradise, most individual accounts go crazy and only the overall balance – the gross national product of the vacuum, as it were – makes sense. The properties of the vacuum and its symmetries show up in the way in which a force depends on the distance over which it acts.

Let us first consider the electromagnetic force at long range. It derives from the group U(1) in the electroweak product group SU(2)⊗U(1). Because U(1) is one-dimensional, there is only one velpon, the photon. All one-dimensional groups are Abelian, so the photon must be a simple beast; in particular, it cannot carry electric charge. It must have spin 1, so the photon is a massless vector boson. Because it is massless, it has infinite range, so the electromagnetic force can extend indefinitely far in space. The behaviour of the force as a function of distance can be inferred from the self-energy picture of a free electron.

The naked electron is dressed at all times in a photon cloud (Fig. 13.7). The electron behaves like a mad gunner, firing off photons at random in all directions. The range of the photon is infinite, so the number of photons per second that pierce a spherical surface centred on the electron is independent of the radius of the sphere. The surface of the sphere, being two-dimensional, increases as the second power of the radius. Thus, because the number of photons over the whole surface is constant, their density

Fig. 13.7 Multi-loop electron path. Because of the active role of the vacuum, the electron is dressed in a photon cloud.

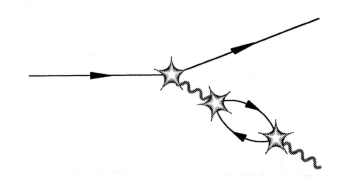

Fig. 13.8 A virtual photon dissociates into an electron–positron pair. The pair, created from the vacuum, changes the propagation properties of other charges nearby.

decreases inversely proportionally to the square of the distance R to the electron. Since the strength of the force is proportional to the probability of exchanging a photon, it scales as the photon density, so that we find: (force) = constant/R^2. This is `Coulomb's law` of static electric attraction. The constant in Coulomb's law is equal to the product of the charges of the interacting particles. It can be positive or negative, depending on whether the charges are equal or opposite in sign.

How does the electromagnetic force behave at short range? At small distances, quantum effects become important; in particular, the behaviour of the photons in the photon cloud around the electron is modified by vacuum polarization processes such as the appearance of a single electron–positron loop (Fig. 13.8). What does this do to the force? Because the force derives from a symmetry, the most probable structures are those that are invariant (that is to say singlets) under that symmetry. In practice, this means that a system 'strives' to be charge-neutral because a system with no net charge (be it electric, weak or colour charge) still looks the same from the outside if all internal charges are swapped by applying a symmetry.

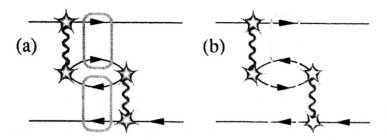

Fig. 13.9 Screening by vacuum polarization. (a) The virtual photon dissociates into an electron–positron pair such that the effective charge density in space increases. In (b) the pair is oriented differently so that the charge density is diminished. The latter case is more neutral (meaning more U(1)-symmetric) than the former and is therefore more probable. This screens the two interacting charges from each other, to a certain extent.

Now let us compare the two diagrams in Fig. 13.9. In the one at (a) the electron–positron pair created by the vacuum is placed in such a way that the electron is closest to the original electron; this creates a region in space-time with charge $-2e$. This is definitely not charge-neutral. Contrariwise, in (b) the positron is nearest the naked electron; this creates a region in space-time with zero charge. Therefore, the right-hand diagram is the more probable: the naked electron attracts positrons which the vacuum spontaneously and temporarily creates in its vicinity.

This shields the bare charge of the electron as seen from the outside: the vacuum provides a screening barrier to those who would observe the naked electron. This preference for electron–positron closeness prompted the invention of the term vacuum polarization. As argued above, it is nothing but the 'striving' of a system for charge neutrality, which in turn is the same as saying that there is an underlying symmetry. The lowest energy state is a singlet under the symmetry in question. Accordingly, all forces that are due to local symmetry are subject to vacuum polarization.

In the renormalization process, the dressed charge of the electron (the bare charge corrected for the shielding effects of the vacuum polarization) was made equal to the observed electron charge. This means that the naked charge was chosen expressly to leave a finite residue when the shielding charge is subtracted. Thus, at large distances, Coulomb's law (which is due to the geometrical dilution of the photon cloud in which the electron is dressed) prevails. But as we move closer in, we experience the usual quantum effect: the influence of the uncertainty laws becomes greater. When the region in which the symmetry tries to establish charge neutrality becomes so small that it approaches the electron's Compton length, the screening of the vacuum polarization begins to lose its efficiency because it becomes more and more uncertain just where in space a process takes place.

As we approach the naked electron, less of the bare electric charge is shielded; therefore, the perceived charge increases, and *the force increases more rapidly than Coulomb's law.* The $1/R^2$ behaviour does not disappear, but the coupling 'constant' in the numerator increases as R becomes smaller. Indeed, because of the divergence of the Feynman diagrams for vacuum polarization, the apparent charge of the electron increases without bounds as we get closer and closer to the naked electron! And yet, in our large-scale world, electric and magnetic forces are not infinitely strong; we owe this to the vacuum, which shields us from disaster.

In the description of a force as being due to the exchange of quantum beanbags, it is fairly easy to see how the recoil of the throw and the impact of the bag can transmit momentum. But we remember that the bag can contain databeans, giving to the exchange a greater complexity and richness than one might at first expect. One of the complications comes from the law of spin and statistics, which produces interference effects that cause repulsion or attraction, depending on the spin of the boson† that is exchanged.

In the case of the non-Abelian groups, the complexity of the interaction is very considerable, so let us restrict ourselves to the one-dimensional U(1) group of electromagnetism and ask about the U(1) velpon: what's in the bag, and how do the contents of this quantum beanbag influence the behaviour of the force on a large scale?

As was shown above, the U(1) velpon is a massless uncharged spin-1 particle, the photon. Thus, it has two allowed polarizations: one with right-handed and one with left-handed helicity. In this particular bag, then, there is one databean, which can transmit one bit of information: an L or an R. Which one it is depends on what happens at the vertex. To see what this implies, consider a straight row of electrons and imagine that there is another electrically charged particle (for example a positron) sitting a little distance from that row. The electrons and the positron are aware of each other's presence due to the constant exchange of photons. Consider two electrons on the row (Fig. 13.10) at equal distances from the positron. If

---

† It is not easy to explain the details, so I will just quote the results. If s is the spin of the exchanged boson, then the force is always attractive if s is zero or even. Thus, the force of gravity, mediated by the graviton which has $s = 2$, is exclusively attractive. But when s is odd, the force can be repulsive or attractive, depending on the sign of the charge. Thus, the electric force, mediated by the $s = 1$ photon, can repel or attract. However, these rules are only valid for *static* interaction. To make things even more complicated, some of the interactions occur only with *moving* charges (e.g. the magnetic force). Such generalized magnetic forces can have rather bizarre consequences, and are almost impossible to visualize.

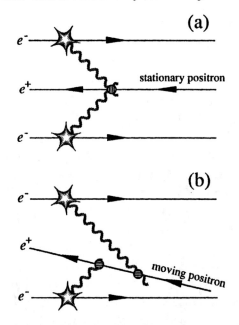

Fig. 13.10 (a) A positron is standing still with respect to two electrons. Photons emitted simultaneously by the electrons meet the positron at exactly one point. (b) But if the positron moves, the photons are no longer received simultaneously. Compare with Fig. 8.5.

each of these fires off a photon simultaneously, the photons will arrive at the positron at the same time. Because of the symmetry of this situation, the two photons have, on average, equal and opposite effects. Thus, nothing happens to the positron.

Now imagine that the positron is moving (Fig. 13.10). As we saw in our discussion of Lorentz symmetry and relativity, the simultaneously emitted photons then do not arrive at the same time, just as we saw Laurel tip his hat before Hardy. Accordingly, the effects of the two photons no longer cancel and the positron experiences a net transfer of momentum: it feels a force. Notice, by the way, that this is purely a relativistic effect because of the lack of a uniquely defined time order of events. The faster the electrons move, the more serious the timing difference and the stronger the force.

Can we identify this process? Yes: we are dealing with the `magnetic force`. The electrons moving with respect to the positron produce a `current`, and stronger currents (i.e. faster and/or more moving charges) induce stronger magnetic fields. The fact that the photon is a vector particle is evident in the vector character of the magnetic field: it is perpendicular to the direction of the current and perpendicular to the direction towards the current. The fact that, on average, there are as many L-photons as R-photons in the combined field of the electrons implies that the magnetic field must obey an additional constraint. It turns out that this constraint obliges the magnetic field lines to be closed loops; there are no loose ends or magnetic charges anywhere.

Arguing along the same lines as above, we would expect that a charged particle with spin behaves, in some sense, like a ring current and must therefore have a magnetic field. This turns out to be the case: an electron has a magnetic moment, a quantum unit of magnetic field strength. If one accepts this reasoning, it would at first seem bizarre that the neutron is also observed to have a magnetic moment. After all, how could a neutral particle act like a ring of rotating charge? We are bailed out here by the fact that the neutron is composite: the electromagnetic behaviour of the neutron is due to its *substructure* (Fig. 13.11).

In the old strong-interaction picture, one would have to argue that the neutron temporarily splits into a proton and a pion, each of which could feel the photon. When one calculates the magnetic moment in this way, the results do not match the observations. In the quark picture, the photon can interact with any of the constituent quarks inside the neutron. This is presumed to produce the correct magnetic moment, but the calculations are very difficult.

So much for the magnetic force. But we are dealing with *electro*magnetism, so one may wonder if the Coulomb force is also related to the polarization of the virtual photons. We have used up the degrees of freedom described by the L- and R-polarizations, so it would seem that our beanbag cannot carry enough databeans to transmit the electric force as well. But here the uncertainty principle comes the the rescue. A free massless particle cannot have a spin component perpendicular to its direction of motion. But the photon that is exchanged in a Feynman diagram is a *virtual* photon, so it is indeterminate whether it has mass or not! Thus, there is a finite probability that the virtual photon does have transverse polarization. In this guise, the photon can have two extra degrees of freedom. One is called a time-like photon, the other is called† a longitudinal photon. Because it does not have a handedness, the

---

† The name may be confusing. It derives from a particular way of describing the polarization, called the Coulomb gauge, in which the photon is linearly polarized perpendicular to its direction of motion. In that gauge, the longitudinal polarization is forbidden. In the radiation gauge, we would call this a 'transverse helicity' photon.

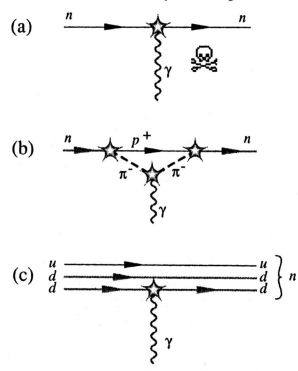

Fig. 13.11 (a) The direct interaction between a photon and a neutron is not allowed because the neutron is electrically neutral. (b) Taking the strong interaction into account, a neutron can temporarily dissociate into a proton and a π^--meson. Because these particles are electrically charged, the photon can interact with them. (c) In the quark picture, the photon can interact directly with any one of the quarks inside the neutron.

force transmitted by the time-like and longitudinal photons is a scalar: it is the electrostatic force, the Coulomb force pointing directly at a charged particle. At short range, the effects of the forbidden polarizations of the virtual photon are noticeable; at very large distances, the effect of the time-like degree of freedom cancels against that of the longitudinal degree of freedom. A long-range photon (a real photon, as seen by your eyes when you are reading this book) has only two possible helicities.

Thus, electricity and magnetism are intertwined because of Lorentz symmetry. The large-scale behaviour of the electric and magnetic forces are due to the photon's spin, to relativity and to quantum uncertainty. Analogous effects exist for all other quantum forces, but they are much more complicated.

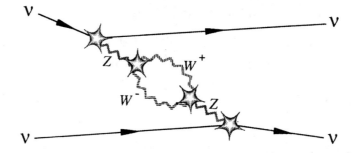

Fig. 13.12 A virtual Z-boson dissociates into a $W^+ - W^-$ pair. This weak vacuum polarization is analogous to the temporary creation of an electron–positron pair out of a propagating photon; see Fig. 13.8.

13.5 The weak force

Let us now consider the weak force. It derives from the group SU(2) in the electroweak product group SU(2)⊗U(1). There are three velpons, all spin-1 bosons: W^+, Z and W^-. They have masses of about 90 GeV due to the symmetry breaking that is built into the unification of U(1) and SU(2). Because SU(2) is non-Abelian, the W-bosons carry weak charge. They also have mass, so at distances beyond the Compton length of the W- and Z-bosons the force cuts off exponentially. At distances smaller than that, the force obeys Coulomb's law; this combination of Coulomb-plus-cutoff is called Yukawa's law.

The mass of the W and Z prevents the weak force from extending beyond their Compton length (about a hundredth of the size of a proton); therefore, we need not bother about weak charge screening at large distances. At small distances, we experience the usual problems with infinite amplitudes for the Feynman diagrams of self-energy and vacuum polarization (Fig. 13.12). But the electroweak theory was constructed with massless gauge bosons that acquired a mass via spontaneous symmetry breaking. The fact that the massive weak velpon was made by joining the degrees of freedom of massless particles gives hope for renormalization. As it happens, this hope is fulfilled: the electroweak theory can be renormalized, so self-energy and vacuum polarization are under control.

Just as in the case of the naked electron, vacuum screening at close range becomes unimportant. But the way in which the effective weak charge behaves at close range is not as simple as in the case of the electric charge. This is due to the fact that SU(2) is non-Abelian, so that some gauge velpons (namely W^+ and W^-) carry weak charge. Thus, there exist vacuum polarization diagrams of the type shown in Fig. 13.12. Now the weak charge of the electron is $-\frac{1}{2}$, of the neutrino $+\frac{1}{2}$, of the Z-boson 0 and of the W^{+-}-bosons it is ± 1. Therefore, even though the vacuum polarization brings v and W^- or e^- and W^+ close together, the weak charge of the W is not fully compensated. As a result, the effective weak charge of the fermion is smeared out in space. Naturally, as we probe deeper into the smeared-out region, we leave more and more weak charge behind us, so that the effective charge diminishes and the weak coupling constant decreases somewhat. This means that the weak force increases less quickly, when the interacting particles get closer together, than Coulomb's law would indicate.

13.6 The colour force

Finally, let us consider the colour force. It derives from the group SU(3), so there are eight spin-1 massless velpons, called gluons. Because SU(3) is non-Abelian, some of the gluons (six, in fact) carry colour charge and the others are colour neutral. There are three colours, arbitrarily labelled red (R), green (G) and blue (B). A colour combination of equal parts R, G and B is said to be white or colour neutral; combinations of equal parts colour and anticolour are white, too. Any white state is symmetric (a singlet) under the SU(3) swapping of colours; the most tightly bound states are white. This is the colour equivalent of 'opposite electric charges attract each other'.

The gluons that mediate the colour force are massless, so we might expect it to have infinite range and otherwise to behave like the electromagnetic force, too: after all, the photon is a massless vector boson, just like the gluons. But it turns out that the colour force at long range does not drop off like Coulomb's inverse-square law. On the contrary, there are indications that it is independent of distance, or at any rate changes very slowly with distance. Contrariwise, at short range the effective colour charge drops rapidly. This dramatic difference between the colour force and the electromagnetic force is entirely due to the vacuum polarization. In an SU(N) force there are $N(N-1)$ charged velpons, which tend to smear out the effective charge of a naked particle at close range, and N fermion–antifermion pairs, which tend to screen the charge and thereby require the bare charge to increase. When $N = 3$, one expects that smearing effects win: six coloured gluons against three quark–antiquark pairs. Indeed, calculations indicate that the colour force drops rapidly as the distance over which it acts shrinks to zero. This remarkable phenomenon is called `asymptotic freedom`.

Of course things are not quite as simple as sketched here; in fact, this behaviour is very counter-intuitive because it is due to the 'magnetic' part of the colour force instead of the 'Coulomb' part. Just remember how

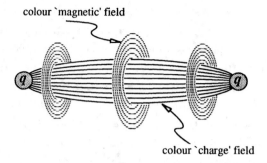

colour `magnetic' field

colour `charge' field

Fig. 13.13 Colour field string: the quarks are colour-charged, and the colour force between them can be represented by field lines, analogous to electric field lines. Similarly, there are colour-magnetic field lines. When the colour charges are pulled apart, work must be done against the force. This intensifies the field; it is the colour equivalent of the induction of a magnetic field by moving electric charge.

oddly a regular (electromagnetic) magnet behaves, and then consider how complicated things can become when we are dealing with the combined effect of eight gluons instead of a single photon!

If quarks are slammed together at extremely high energies, so that they are forced very close together, their colour interaction becomes quite mild. Thus, mesons and baryons cannot be too small, or else the colour force would be insufficient to bind them together. It is encouraging that theoretical calculations in this almost-free domain agree with the data obtained in high-energy accelerators.

At large distances, the opposite happens (Fig. 13.13). Because the colour-carrying gluons are so numerous, the effective charge on the dressed quark increases, even as the geometric Coulomb dilution tries to lower the strength of the force. The overall result is that at large distances the colour force changes only slowly; in fact, it may even become constant. Apparently, the structure of the vacuum tablecloth modifies the wrinkles caused by a local gauge twist. The way in which the force field is modified is such that the wrinkles, instead of spreading radially, become confined in one direction only, forming a tight bundle that is practically one-dimensional. We might say that the vacuum polarization confines the colour field lines to a narrow tube or string.

For people who like analogies in physics, it may be interesting that this bunching of colour field lines is analogous to the Meissner effect, in which a magnetic field in a superconductor is pressed together into a narrow flux tube. If the field tries to spread, it induces a huge electric current in the infinitely conducting surrounding medium. This produces a reaction force on the magnetic field to compress it back to where it was. Similarly, the presence of copious quantities of gluon–antigluon pairs makes the vacuum behave as a superconductor of colour-magnetic current. If the colour field tries to spread, it induces a reaction force due to the gluons in the vacuum that pushes the field back; the colour field lines are almost literally glued together.

The amount of energy it takes to increase the distance between two quarks is, just as in classical mechanics, equal to the force between them multiplied by the increase in their distance. And since the force between quarks is almost constant at large distances, it costs a prodigious amount of energy to pull them apart (Fig. 13.14). Instead of two separate quarks, each surrounded by a spray of field lines, we obtain *four* quarks, pairwise connected by a bundle of field. This is why free quarks have never been seen, but always form bound states (mesons and baryons) that are white. Due to the existence of so many colour-charged gluons, the penalty for deviation from colour neutrality is especially severe and whiteness is the order of the day. Because of vacuum polarization the strength of the colour interaction (and hence of the strong force) no longer depends on the colour coupling constant being large. As long as there are many different coloured gluons, the force can be very strong at long distances, even though the coupling constant (the colour equivalent of the electric charge e) is not itself large.

When two quarks are forcibly pulled apart (for example because they receive a high-energy knock in an accelerator), the energy E used in the pulling is stored in the colour field. When E exceeds the mass equivalent mc^2 of a quark–antiquark pair, the field lines snap and their ends are capped by a quark and an antiquark (there is enough energy stored in the field lines to allow these to form). In this way, a high-energy collision with a hadron (a bound-quark state) produces a spray of mesons and baryons, instead of a shower of free quarks. But the initial direction that the quarks had, before their field lines were stretched to breaking point, can still be recognized in high-energy experiments: the hadron sprays are often bunched together in narrow `jets`, indicating the direction of motion of their ancestral quark (or sometimes gluon).

The above does not, of course, mean that all is moonlight and roses in this subject. The moonlight is made of photons, which we understand pretty well; but the roses contain atomic nuclei, made up of nucleons in which the colour force is hidden. The precise behaviour of that force cannot yet be calculated, even on the most powerful computers, when the distances between the quarks become large. The gross contributions from quark–

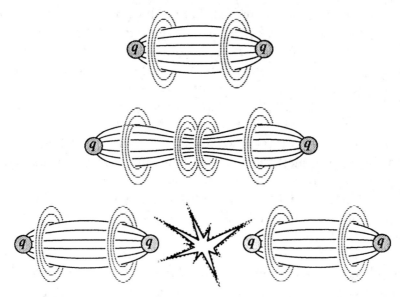

Fig. 13.14 New baryons can form when the string of colour field lines snaps. When the colour charges move apart, the resulting induction strengthens the colour field. The energy stored in the field can become large enough to make a quark–antiquark pair because of the equivalence of mass and energy.

antiquark and gluon–antigluon pairs in the vacuum have been computed, but in the whole pack of forces there are quite a few jokers. For example, the electroweak field can generate `magnetic monopoles`, which carry a single magnetic charge comparable to, say, a lone magnetic North pole. They are the magnetic equivalent of electric point particles such as the electron and the positron. If these things exist in our Universe, they can cause some wild processes, even at low energies.

For example, a magnetic monopole could attract the *uud*-quarks that make up a proton, confine them very briefly to a very small region, and instantly convert them into a π^0 and an e^+; or a cloud of magnetic monopoles might encounter the Solar System, short-circuiting the magnetic fields of the Sun and the planets, so that the fields would disappear in a flash! Other unsolved problems arise because of the symmetry breaking in SU(2)⊗U(1): do the Higgs particles exist as free objects? If we must take them seriously, does the Higgs field contribute to the vacuum polarization? Many such problems remain, and undoubtedly the present picture of forces is only an approximation of what it will be a decade from now.

14

Excelsior! The ascent to SU(∞)

14.1 Motives for unification

A snowflake is symmetric, but that does not teach us much about the properties of frozen water. All we observe is that a snowflake has an axis about which it is symmetric under rotations over 60 degrees. We are left guessing about other interesting properties of snowflakes: melting point (it is not 60 degrees!), heat capacity, chemical composition and so forth. But the discussions in the preceding chapters show that the symmetries of the vacuum (Lorentz, U(1), SU(2) and SU(3)) are awfully restrictive and leave precious little room for manoeuvering.

The standard model of these four forces accounts very well for the properties of the observed world. The U(1) theory of quantum electrodynamics is a particularly amazing success, and yet there are some free quantities left. The most nagging of these are the values of the masses of the leptons and quarks; it is especially galling that the masses in the various generations do show a pattern of systematic increase, but it is not orderly enough to allow us to read the underlying mechanism. Also, the extent to which some particles can be superposed into quantum chimaeras (as expressed by the mixing angles) must be derived from experiments. Finally, it is observed that electric charge is quantized and that quarks carry a rational fraction of the electric charge of the electron, even though this is not a necessary outcome of the theories we have encountered so far.

Clearly, *we want to be more restricted, not less*. And if we retain the most productive idea so far, namely that forces are generated by local symmetries, it is natural to inquire if there might not be a master symmetry from which all forces can be derived. The quest for such a symmetry is historically associated with the blunderbuss expression `unification`. We need not be

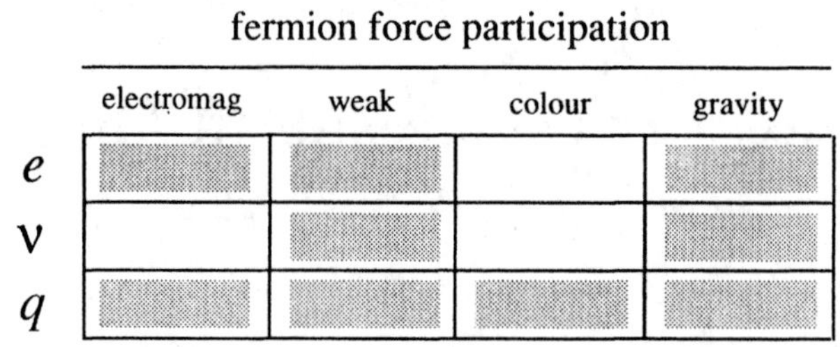

fermion force participation

	electromag	weak	colour	gravity
e	▨	▨		▨
v		▨		▨
q	▨	▨	▨	▨

Fig. 14.1 Grouping of the three basic fermions according to the forces they feel. A grey box means that the particle on the left (electron e, neutrino v or quark q) feels the corresponding force listed in the top row.

unduly impressed by the results obtained so far, nor by the advertisements for the various competing models, because the subject has such a phenomenal historical importance that theorists attempt to unify (except among themselves) at the slightest provocation.

Nevertheless, there is one working example of unification which we have encountered already: it is the case of the electroweak force, which was built from the group SU(2)⊗U(1). This isn't quite the real thing, though, because we are not dealing with one whole group, but with a product group. This leaves an arbitrary parameter which prescribes to what extent the neutral vector bosons of the U(1) and SU(2) groups are mixed up. Also, the group is spontaneously broken, giving a mass to its velpons. To some, this may be an indication that SU(2)⊗U(1) is not fundamental enough, and is a residue of a deeper symmetry. In that case, we must consider the W- and Z-particles (and, who knows, maybe even the photon!) as composites. There are precedents for this, as in the case of the strong nuclear force, which turned out to be a residue of the colour force. But it is surely encouraging that some sort of unification can be achieved at all; moreover, there are other good reasons and hints that unification is not hopeless.

First, the basic interactions overlap to a certain extent (Fig. 14.1). The quarks are subject to the colour, the electromagnetic and the weak forces. In addition, some leptons feel the electromagnetic and the weak forces; the remaining leptons are subject to the weak force only and all particles should respond to gravity. Thus, one can traverse the whole world of forces by going from one fermion family to another.

Second, the quarks have electric charges that are whole thirds of the unit lepton electric charge. A suitable symmetry of leptons and quarks could produce this quantization, which is otherwise unexplained.

Third, the behaviour of the coupling constants of the forces is very suggestive (Fig. 14.2). As we have seen, due to vacuum polarization the electromagnetic coupling becomes stronger as we bring particles closer together; the weak coupling decreases somewhat; and the colour coupling drops very steeply as the distance between the interacting particles is decreased. These trends, of course, reflect the increase in the number of charged velpons that occurs when the dimension N of SU(N) increases, so it might not indicate anything special. However, calculations show that the coupling constants are approximately the same at a distance of 10^{-31} metres (if this were the Compton length of a particle, its mass would be about 10^{15} proton masses). Unless this is a cruel coincidence, we can surmise that on these supersmall scales all three forces are the same, indicating the existence of an underlying grand symmetry.

Fourth, the existence of several generations of leptons and quarks is a broad hint that there is more symmetry than meets the eye. The table of properties of the fundamental (beware that word!) fermions reveals not much more than that for each lepton doublet there is a quark doublet and that the doublets increase in mass from one generation to the next. Even though no strict regularity is apparent, it would be foolish to ignore this signpost.

Fifth, the force of gravity has so far stubbornly resisted all attempts at bringing it into the quantum fold. This is totally unacceptable and perhaps a deeper symmetry would allow gravity to be brought to heel. In that respect, it is intriguing that all symmetries are some kind of rotation. Lorentz symmetry is a peculiar type of rotation of space-time; the other three (U(1), SU(2) and SU(3)) are rotations in some internal space. We know of no compelling reason why the mathematical behaviour of symmetries with such disparate origins should be the same.† Maybe this similarity is a broad hint of a unity between gravity (due to local Lorentz symmetry) and the other forces.

Sixth, there is a very vague hint in the way in which relativity and quantization mesh together. There may be an underlying unity in relativity and quantization. It is then plausible to expect that their characteristic constants, c and $\hbar$, are somehow related to each other. The numerical values of c and $\hbar$ are surely unimportant, for these depend entirely on the physical units we use in their measurement. In fact, we could cheerfully adjust our measuring units such that $c = \hbar = 1$ and indeed this is often done: to the professional physicist, keeping track of c and $\hbar$ is a complete waste of time. Since the values of these constants are irrelevant, any significant relationship

---

† Almost the same, that is. The Lorentz group is not 'compact' (an esoteric mathematical property), whereas the other three are.

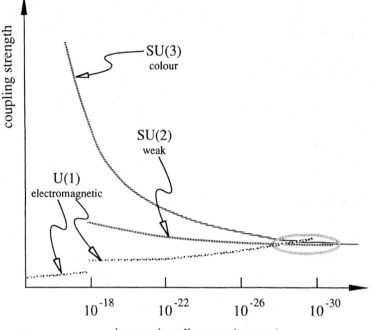

Fig. 14.2 Force strength versus range. As we probe smaller and smaller distances, the electromagnetic force due to the U(1) symmetry increases its effective strength. At the point where the weak force begins to exert its influence (at about 10^{-18} metres) the unification of these forces causes a slight jump in the electromagnetic strength. The SU(2) weak force becomes a little weaker, and the SU(3) colour force becomes much weaker at smaller distances. In the neighbourhood of 10^{-31} metres the strengths of the forces are similar.

between them must be expressed as a dimensionless number, that is, a purely mathematical number (such as 42, π, or $\sqrt{2}$) independent of the units of measurement.

The dimension of Planck's constant is, as we have already seen, that of angular momentum: $[\hbar] =$ mass $\times$ speed $\times$ length $=$ energy $\times$ time. The dimension of the speed of light is, of course, $[c] =$ speed $=$ length/time. Because Planck's constant contains the dimension of mass and c does not, it is impossible to concoct a dimensionless number from these two. We need at least a third ingredient. There is such an ingredient, besides relativity and quantization, in gauge theories of the forces: namely, a symmetry group. As we have seen so often, a symmetry always leaves something unchanged; that means that each symmetry has an associated conserved quantity. In the case of local symmetry groups, that quantity is the charge. We now recall Coulomb's law, which was seen to be a consequence of the U(1) symmetry of electromagnetism: (force) = (charge)^2/(length)^2. Remembering also the old truth that the product of a force and a displacement is an energy, we can rewrite Coulomb's law in dimensional form as (charge)^2 = (energy) $\times$ (length). If we call the charge e, we see immediately that one, and only one, dimensionless number can be constructed from $\hbar$, c and e, namely $\alpha = e^2/\hbar c$. If we insert the experimentally observed values of the constants, we obtain the dimensionless number $\alpha = 1/137.036$.

This number, the electromagnetic `fine structure constant`, plays a very important role in quantum theories. For example, the energy E of the most stable form of the hydrogen atom is given by $E = \frac{1}{2}mc^2\alpha^2$, where m is the mass of the electron. An analogous dimensionless number could be derived for the other forces in Nature as well, but then we would have to account for four dimensionless numbers, which is much more than we should like: one number would do fine to summarize the almost miraculous cooperation between relativity, quantization and symmetry. The desire to make do with just one dimensionless number is an incentive to prove that all coupling strengths in Nature are variations on the theme 'e'. In other words, there should be one supergroup, of which all others we have encountered are offspring. We would still have to worry about the origin and value of α. For your next Nobel prize, maybe you should start thinking about that.

14.2 Grand unification

Clearly, the motivation for striving towards some grand unification is strong, but from here on we must grope in the dark: much of the remainder of this chapter is pure speculation. The forces we know today are remarkably well described by the `standard model` of the SU(2)$\otimes$U(1) and SU(3) groups (plus the local Lorentz group for the non-quantum theory of gravity); our earlier efforts leave scant observational data unaccounted for, whereas we need unexplained observations to point the way ahead. The very successes of the present theories have removed all clear indications of the road to follow. Accordingly, the professional physics journals are cracking at the seams with grand unification proposals, which – depending on the preference or prejudice of the reader – range from the ridiculous to the sublime. From this embarrassment of riches I will present just one example, not because it is entirely right (we know from the most recent observational data that it needs modification), but because it is wonderfully simple, is a natural extension of our preceding discussion and illustrates the general idea of a grand unified theory, abbreviated GUT, quite well. The model is based on the group SU(5).

We are looking for the smallest entire group (not a product group) of which the known quantum symmetries of Nature (U(1), SU(2) and SU(3)) are subgroups. Furthermore, the fact that there is a quark doublet for every lepton doublet, with more or less regularly increasing masses, gives us a broad hint that in some sense a lepton is like a quark. This means that we could try to lump leptons and quarks together in one fermion multiplet. Lest this seems an outrageous step, consider the fact that the electron and its neutrino have wildly different properties, and yet the weak interactions can be understood by grouping these particles together in an SU(2) doublet.

Because the grand unification group must contain SU(3) as a subgroup, the grand fermion multiplet must contain at least one quark of each colour R, G and B. Because SU(2) must be a subgroup, too, the multiplet must also contain two weak fermions; the group U(1) can be a subgroup automatically if one of these fermions is electrically charged. All this suggests that we use $3 + 2 = 5$ as the dimension of our multiplet and SU(5) as its symmetry group.

In the SU(5) model, the fermions of each generation are arranged in two multiplets; the first generation consists of the quintuplet $(\bar{d}, v, e^-)$ and the decuplet $(\bar{u}, e^+, u, d)$. When counting members, recall that each quark comes in the three colours R, G, B, which gives five for the first and ten for the second multiplet. The second generation is constructed likewise with the c- and s-quarks, the muon μ^- and its neutrino v_μ, while the third generation contains the t- and b-quarks, the tau lepton τ^- and its neutrino v_τ.

A few remarkable things follow directly from this grouping. First, quarks of the same flavour (e.g. u_R and u_B) can interact by the exchange of gluons, just as before. Thus, of the $5^2 - 1 = 24$ velpons generated by the SU(5) symmetry, eight are old friends. Second, because the leptons can interact via the electroweak component of the SU(5) force, four more velpons are accounted for: they are the W^+, Z, W^- and the photon. Thus, 12 new velpons – called X-bosons in this case – should exist. Third, because U(1), SU(2) and SU(3) are subgroups, all charges in this model should be rational combinations of 1, 2 and 3. That is, the charges of whatever kind carried by velpons or fermions can only be (any product of 1, 2 or 3) divided by (any product of 1, 2 or 3). Examples are $\frac{1}{6}, \frac{1}{2}, \frac{2}{3}, 1, \frac{4}{3}, \frac{3}{2}$ and the like. This is exactly what is observed; for the first time, we can understand the quantization of electric charge and can thereby explain why the charge on the electron is the same (except for its sign) as that on the proton.

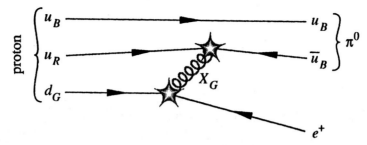

Fig. 14.3 Proton decay diagram $uud \rightarrow e^+ + \pi^0$. A d-quark changes into a positron by emitting a hypothetical X-boson. This new type of velpon is absorbed by a u-quark, which changes colour and becomes a $\bar{u}$. This then forms a neutral pion with the nearby u. Because grand unification puts leptons and quarks together in the same multiplet, things like this must happen in all such theories. In other words, matter as we know it is unstable in grand unified theories, because these theories presume that an electron is somewhat like a quark.

Other features, however, are less attractive. For example, the SU(5) symmetry must be broken in a bizarre way: once to give mass to the W and Z, and once to give mass to the X-bosons. Especially the latter should have a large mass: if they did not, quarks could readily turn into leptons, thereby destroying the hadrons in which they are bound, for example by the process of Fig. 14.3, in which a proton decays into a pion and a positron. This means that *all matter is unstable*; because in any truly unified theory the SU(3) group must be a subgroup, instead of being a factor in a product group, the decay of ordinary matter is a general feature.

In any GUT, all protons pop; the only question is, what is the probability for that: what is the proton's half-life? In the above process the positron would soon disappear by capturing a stray electron, and the pion would decay, too; thus, the proton would vanish in a blaze of photons, carrying the full GeV of the proton's rest mass energy. Such gamma rays are quite lethal, but we know that people do not typically die from the self-destruction of the protons in their bodies. An average person contains some 10^{29} protons. Using some standard numbers on the lethality of radiation doses, we can say conservatively that the half-life of the proton is at least 10^{17} years, ten million times the present age of the Universe; if it were less, our own protons would kill us. Current experiments come up with lifetimes that are much larger still: about 10^{32} years. Accordingly, the Compton length of the X-boson must be very, very much smaller than that of the proton: 10^{-31} metres or less. This X-boson must be exceedingly massive: at least 10^{15} proton masses.

In order to obtain such a large mass, the SU(5) symmetry must be broken with a vengeance. This makes many people uneasy because it requires the addition of all sorts of Higgs particles, mixing angles and masses. These purely external requirements clutter the theory and are diametrically opposed to our desire to be more restricted instead of less. On the other hand, the Compton length of the X-boson rings a bell: where did we see that number before? Well, we noticed above that the coupling strengths of the three quantum forces may be very closely similar in this regime. This gives an extra indication that something significant is happening there.

14.3 Supersymmetry

Many things are unaccounted for in today's GUT models. For example, it is not explained why there is more than one fermion multiplet. Who ordered six of them? Indeed, there may be more. Their multiplicity seems to hint at another symmetry somewhere, but marching from SU(5) to SU(∞) in order to include it is self-defeating. The number of velpons increases as $N^2 - 1$, so we would have to add more and more external quantities to orchestrate their behaviour.

Possibly, we are on the wrong track if we try to increase the dimension of our group. Instead, we might try to increase the number of dimensions of the stuff that the symmetry works on: the vacuum. The extension of the three dimensions of space to the four dimensions of space-time proved to be very restricting and very rewarding: the Lorentz symmetry of space-time forbids a lot, but at the same time imposes a wealth of delicate structure. Perhaps the abstract dimensions in which the SU-rotations take place are not so abstract after all. Quite possibly, these degrees of freedom have something in common with those that occur in our four-dimensional space-time. It would be premature to jump to the conclusion that we can move in the direction of these extra dimensions; in fact, we cannot even freely move in time.

To illustrate this point, let us imagine that there are beings that live in a parallel universe where the abstract dimensions of, say, SU(3) are as ordinary as spatial directions are in our Universe. Suppose that, in such a world, space-time as we know it is unheard of, but that these SU-beings had discovered that particles have spin. Undoubtedly, they would ascribe this quantum number to a degree of freedom in an abstract dimension. They might even be able to deduce the equations of motion in space-time from the observed properties of spin. Likewise, it is not unthinkable to attempt something similar in our Universe. In a way, we were doing this all along: with each space-time point we associated a Yang–Mills type 'rotation' about some abstract 'direction'. All we do is to take this a bit more literally. Such an approach is a natural extension of the historical trend indicated in the preceding chapters. From 'particles and forces in space and time' we went to 'particles and forces in space-time', to 'quanta in space-time' and to 'quanta in vacuum' ... why not just 'vacuum?' Perhaps the existence of so many quanta with different charges and masses is trying to tell us that what we observe is due to an as yet uncharted motion in a higher-dimensional space.

Another reason to suspect an underlying multidimensionality is that when we descend the distance ladder deep into the realm of the exceedingly small, we do *not* find that things become simpler. The uncertainty relations might lead to the expectation that quantum states become increasingly simple as their extent in space becomes smaller. But the quantum beanbags, instead of containing fewer and fewer databeans, become more and more complex because the dimension of their symmetry group keeps increasing! Nobody yet knows what this baffling behaviour is trying to tell us about Nature. It almost seems as if, by squeezing a quantum in ordinary space, we spread it out in the direction of some unknown dimension, preventing us from reaching a scale of smallness where simplicity reigns.

The curious thing is that the equivalence of matter and energy makes it possible for a quantum to gain access to more and more exotic states (Fig. 14.4). The uncertainty relation $\Delta E \, \Delta t > \hbar$, together with $E = mc^2$, implies that any mass state m should be temporarily accessible for times smaller than $\hbar/mc^2$. A far as we know, there is no limit to the smallness of Δt and the mass spectrum of particles continues indefinitely, so that squeezing Δt opens the way to an endless realm of exotic particles.

By the way, the quark masses are really not known very well at all because there's no such thing as a free quark. What would the mass be anyway? Context is all: as we saw in the case of the renormalization of the electron, mass is determined by environment. Of course you can then ask: what environmental effects determine the masses of the other particles? That is, of course, a 500 000-dollar-question (or $ 250 000 if you have to share the Nobel Prize with another clever and lucky person).

We have precious little guidance on how to proceed, but there is at least one property of all the forces discussed so far that is most annoying, namely the fact that we always had to start by guessing at their 'fundamental' multiplets. This is blatantly contrary to the spirit of the enterprise, which is to treat all quanta equally. Starting out with such multiplets amounts to fermion favouritism, surely a leftover from our large-scale preference to treat matter as something real and forces as something ethereal. But now that we know the law of spin and statistics, we can emancipate ourselves

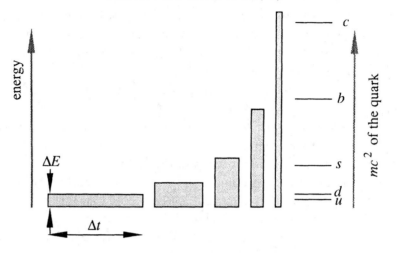

Fig. 14.4 Because of the uncertainty relation, diminishing the residence time Δt of a particle increases its energy uncertainty ΔE. We can represent this product of uncertainties by a rectangle of constant surface. The horizontal size of the rectangle corresponds to the residence time, the vertical direction indicates the energy. Rectangles with constant surface squeezed in one direction must push out along the other axis. Thus, eventually *all* energies (or masses) become accessible. Plotted along the right edge is the mass spectrum of some known quarks.

from this prejudice. If we do, then maybe we can begin to understand such mysterious things as the masses of leptons and quarks. Only a theory that includes fermions and bosons on the same footing can begin to aspire to being a 'theory of everything'.

When we speculate along these lines, we notice that the spin is some sort of rotational charge, associated with the spatial rotation group in a manner analogous to the association between, say, colour and SU(3). We have cheerfully lumped together particles with different charges and masses in the same multiplet before; so why not do that now? Indeed, the fact that fermions can be virtual particles, for example in Compton scattering, might be a hint that we can include them with the velpons in one supermultiplet belonging to some undiscovered supersymmetry.

Why not, indeed? Because the mixing of fermions and bosons is, in a quantum theory, just about the most un-kosher thing you can do. Let us recall how the two classes of spins came about. A particle can be stable only if it is in phase with itself: the phase wheel rolling along the equator of the particle must make mudprints that coincide after the wheel has traced out the particle's circumference (Fig. 14.5). From this requirement we found that

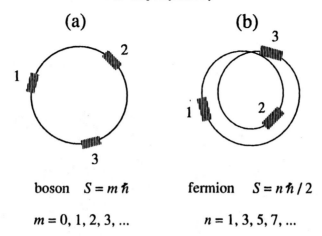

boson $S = m\hbar$ fermion $S = n\hbar / 2$

$m = 0, 1, 2, 3, ...$ $n = 1, 3, 5, 7, ...$

Fig. 14.5 (a) $S = m\hbar$-boson in the case where $m = 3$. (b) $S = n\hbar/2$-fermion in the case where $n = 3$.

there are two classes of particles, namely those where the phase path closes after one loop and those where it rolls around twice before closing on the point where it began.

The first class, bosons, have amplitudes that remain the same if they are rotated by 360°; the second class, fermions, have amplitudes that change sign on every full turn, so that they require 720° to regain their original amplitude. From those differences it was deduced that bosons are gregarious, whereas fermions show Pauli intolerance. Finally, we saw that one cannot transmute fermions into bosons because of the different *topology* of their phase paths: the crossover in fermions cannot be removed unless we lift the phase path into another dimension (Fig. 14.6).

This is exactly what has been done in some recent theories: namely, extend the Lorentz symmetry of space-time to a similar symmetry of a space with more than four dimensions. By demanding the right to step out into another direction, we can twist and untwist the phase paths of bosons and fermions. This allows us to change angular momentum by *half* a unit, so that $\Delta s = \pm\frac{1}{2}$ rather than the $\Delta s = \pm 1$ that was mandatory until now. The extended form of the Lorentz group is called the super-Poincaré group and the corresponding symmetry has become known as supersymmetry.

By means of the super-Poincaré group, we generate particles in fermion–boson pairs. Thus, every classical fermion acquires a supersymmetric boson partner. These hypothetical objects are given the name of the corresponding fermion, preceded by an 's' (guess why!) Thus, the electron and the quark

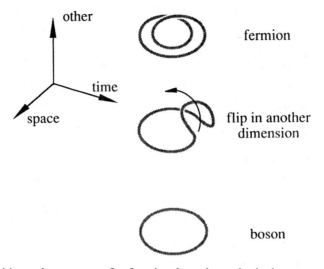

Fig. 14.6 Making a boson out of a fermion by a hypothetical symmetry operation. This is analogous to the untwisting of a fermion phase loop.

have supersymmetric boson partners in the selectron and the squark. Similarly, the supersymmetric partners of known bosons are given the name of their parent particle with the ending '-on' replaced by '-ino'. The photon and the gluon are then accompanied by the photino and the gluino.

What is to be gained by this, other than a way to include hypothetical particles in hypothetical supermultiplets? It would be one thing if the superpartners could be related to known particles, for example by having the gluino be identical with the neutrino, but that does not seem to be the case. Frankly, we are not sure yet of the full potential of this approach, but one possible advantage is a better control over the renormalization infinities that plague theories that are not supersymmetric.

The point is that Feynman diagrams containing loops did in general have infinite amplitude. In quantum electrodynamics, these were tamed by renormalization, but that remains a somewhat awkward procedure, and in any case it fails miserably in all forms of quantum gravity that are straightforward equivalents of QED. Now bosons and fermions differ in one crucial aspect: if we swap two indistinguishable fermions in any given process, the direct and the exchange amplitudes are equal and opposite. A superposition of such processes yields zero. Suppose now that we have a Feynman diagram with an infinite amplitude. By judiciously replacing fermions and bosons with sfermions and bosinos, we may produce another diagram that is also infinite, but that cancels the former diagram exactly!

Taking an overall view instead of treating quantum processes piecemeal turned out to give the right answers in quantum electrodynamics; likewise, supersymmetry may produce superrenormalization. For want of a better phrase, the latter is known in the trade as magical cancellation, a rather unfortunate term in a subject already encrusted with slang that is misguided ('elementary' particle), boastful ('grand' unification, 'super'symmetry), opaque ('strange' quark), corny ('charmed' quark), or even mildly obscene ('bare bottom flavour').

Further study shows that we are very restricted in what we can and cannot do in a supersymmetric theory, and restriction is precisely what the doctor ordered. Moreover, there are tantalizing indications that supersymmetry may snare that old dragon of the quantum knights, gravity. Supersymmetry derives from the super-Poincaré group, which ought to include local Lorentz symmetry somewhere. And local Lorentz symmetry, as we saw in the classical case, generates gravity! Theories that use a super-Poincaré group as a *local* symmetry, in the way pioneered by the Yang–Mills theory, are called `supergravity`. None of these theories has yet produced anything that resembles our Universe, but maybe some day one of them will. If this stunt can be pulled successfully, we naturally expect to find that the gravitational charge, i.e. mass, is quantized, potentially solving one of the nastiest of our unwanted freedom problems: the need to put in the particle masses from the outside.

All unification theories are still in the wouldn't-it-be-nice-if stage and phenomenal difficulties remain. Nobody knows what rules the rules, whether the symmetries show any order or progression (e.g. U(1), SU(2), SU(3), SU(5), ...), or indeed, why all symmetries discovered to date behave as multidimensional rotations. Because there are an infinite number of possibilities between SU(3) and SU(∞) it seems a bit distasteful to take, say, SU(217) at random and see if it works. Probably the road of most resistance, via the supersymmetry models, is more rewarding, but formidable obstacles remain.

14.4 The Planck scale

Another indication that we are courting disaster when we try to include both gravity and extremely massive particles in quantum theories comes from the fact that in classical gravitation an object cannot be arbitrarily compact. Associated with any mass M there is a length, the `Schwarzschild radius` R_S, which has the property that a sphere of mass M cannot be smaller than R_S without suffering total gravitational collapse and becoming a black hole that has a horizon with radius R_S. In Chapter 11, I showed the structure of space-time of a black hole with $R_S = 1$ in dimensionless units. Using units with the usual dimensions of mass, length and time, it can be shown that $R_S = 2GM/c^2$; for our Sun, the Schwarzschild radius is 3 kilometres. This is not the place to go into details about general relativity so I will state without proof that if an object is compressed beyond its Schwarzschild radius, its particles will collapse to a point in space (called `singularity`). Around this point, the curvature of space-time is so strong that there is a sphere with radius R_S centred on the singularity, from within which no signals – not even those that move with the speed of light – can reach the space outside the sphere. The singularity-with-sphere is appropriately called a black hole.

The existence of black holes is relevant here because we now have two length scales associated with a particle: its Schwarzschild radius and its Compton length $\lambda_C = \hbar/Mc$. What would happen if these two were equal? In that case, the particle is so compact that it turns into a black hole. That might not be too bad, except that no information about the particle's quantum properties would be discernible: all the data on what the particle is, what symmetries it obeys, and so forth, are caught inside the Schwarzschild radius. A black hole made by piling up neutrinos is indistinguishable from a black hole built from neutrons.

Let us see what mass a quantum particle must have to become a black hole. Putting $\lambda_C = R_S$, we find $\hbar/Mc = 2GM/c^2$, or $M = \sqrt{\hbar c/2G}$. Putting in the numbers for the various constants, we find $M = M_P = 1.54 \times 10^{-8}$ kg, or 10^{19} proton masses. This quantity is called the `Planck mass`. The Compton length associated with it is the `Planck length` $\lambda_P = 2.28 \times 10^{-35}$ metres. If the Planck length were as big as your fingernail, a proton would have a radius of 10 light years – that is just beyond the star Sirius, which is 8.7 light years away from us! As subnuclear particles go, the Planck mass is prodigiously large. An enzyme molecule contains about 100 000 proton masses; one Planck mass is equivalent to 10^{14} enzyme molecules, enough for a respectable bacterium. Even though it is so huge, the Planck mass comes uncomfortably close to the grand unification mass $M_{GUT} = 10^{15}$ proton masses: only a factor 10 000 is needed, which is not a really large safety margin, given the uncertainties in determining M_{GUT}. Indeed, we might hold the view that it is too much of a coincidence that M_P and M_{GUT} are so close; maybe we have stumbled upon the first sign of a mass scale at which gravity is unified with the other three forces.

When confronted with these facts, we can adopt one of a multitude of theoretical approaches, but one inference is fairly clear: we are forced to reconcile gravity with quantum behaviour. What will emerge is still unknown, but there is no evidence that the particles in our Universe are in imminent danger of gravitational collapse. Therefore, it is not too audacious to speculate that Nature has found a way around the dreaded Planck mass. This might occur, for example, if particle masses were quantized in such a way that the mass spectrum had an upper limit which was smaller than M_P.

In fact, analogous limits are commonplace in the quantum world. For example, the energy states of the electron in a hydrogen atom are given by $E_n = mc^2\alpha^2/2n^2$, where n is a whole number; thus, the energy ranges from zero, in steps of increasing size, to $\frac{1}{2}mc^2\alpha^2$, which is the maximum possible. A mass spectrum arranged like this would imply that there are infinitely many possible 'elementary' particles between the most massive known particles (the W- and Z-bosons, about 100 proton masses) and the most massive conjectured particle (the X-boson, 10^{15} proton masses or thereabouts). Indeed, it would be most surprising if there were no structure whatever in this region: the so-called energy desert between W and X spans more than 13 powers of 10 and it is not to be expected that there are no oases in this vast expanse of mass scales.

14.5 Superstrings

Throughout this chapter I have argued that it makes sense to explore the possibility that the vacuum has more dimensions than the four of space-time. It seems at first that this is in direct conflict with one of the basic requirements of an improved theory, namely that we should be more restricted than before. The more firmly a theory is constrained, the firmer its predictions are, and thus the greater its power. We would have more admiration for a theory that says that the mass of the electron *must* be 0.511 MeV, than for a theory in which the electron mass is a free parameter.

How can we insist that extra freedom in the vacuum will produce more restrictions? In general we cannot, but in the case of forces generated by local symmetries there is an interplay between the number of dimensions and the kind of symmetry that is allowed. It turns out that there are very few symmetry groups that can operate on a high-dimensional vacuum; ideally, we should like to have a unique combination, for example a seven-dimensional vacuum with the group SU(5). We can get a qualitative understanding of such restrictions by considering how spatial symmetries work in lower-dimensional space.

All geometrical figures in a one-dimensional world are alike: they are all simply line segments, and they are as symmetric as you could wish. But if we extend our space to two dimensions, restrictions begin to appear. There is still an infinity of symmetrical figures: 3-rotations produce triangles, 4-rotations produce squares, and so forth. But the symmetric objects no longer look alike, not even within the same symmetry group: an equilateral triangle and a regular hexagon are both symmetric under rotations over 120°, but they are not the same. We are even more restricted in three dimensions: whereas we could build an infinity of regular polygons from one-dimensional symmetric building blocks (namely line segments), we can build only very few regular three-dimensional polyhedra from regular two-dimensional polygons. From equilateral triangles we can build the tetrahedron, the octahedron and the icosahedron; from squares we can build the cube; and from pentagons we can build the dodecahedron. Five objects, that is all. Of course, we can combine various polygons, but they produce less symmetric objects.

Having accepted the possibility that higher dimensions and symmetry groups can only rarely be matched in a local-symmetry theory, what would the overall properties of such a theory be? We can obtain some idea by recalling the way in which rotations (or Lorentz transformations, or any other non-Abelian group elements) multiply. In the section on spin, we asked: what is the rotation R which accomplishes in one operation what rotations P and Q do in two moves? Because the rotations form a group, we know that there must be an R such that $P \otimes Q = R$. But we notice when playing with dice that the rotation axis of R does not lie in between the axes of P and Q. For example, if P is a rotation of Earth over 90° about an axis through Amsterdam and Q is a 90° turn about an axis through Rotterdam (Fig. 14.7), then $P \otimes Q$ is *not* a rotation about an axis through a point in between, such as Leiden. (Actually, it is a rotation of nearly 180° about a point to the East of Amsterdam and Rotterdam, close to Utrecht.)

Thus, P and Q generate a rotation that pokes out of the space spanned by their rotation axes. Similarly, we saw that successive Lorentz boosts† do not produce another boost, but instead produce a boost and a rotation. The two Lorentz transformations conspire to produce a rotation, so again we see that successive applications of a symmetry can lead to a stepping out into another degree of freedom.

In supersymmetry theories we find something similar. Suppose that S and T are supersymmetry twists that do not change the direction of the coordinate axes. Then $S \otimes T = U \otimes$(rotation, translation or reflection). Thus, global supersymmetry operations can produce global changes of orientation, but they are not very interesting; whether or not we rotate Earth about an axis through Utrecht, the surface remains the same. However, by letting the supersymmetry operations be *local*, we obtain something much more

† A Lorentz boost transforms the world as seen by one observer into the world as seen by another one with a different velocity, but without rotating the coordinates.

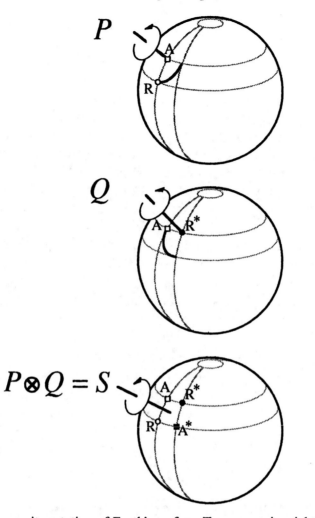

Fig. 14.7 Composite rotation of Earth's surface. Two successive right-angle turns P and Q about two different points produce a composite rotation about an axis which is not a simple combination of the two separate rotation axes. First, we rotate by 90° about point A; this carries point R towards R*. Then we rotate by another 90° about point R*; this carries A to A*. The composite rotation $S = P \otimes Q$, which makes the transformation R → R* and A → A* in a single move, is not a straightforward combination of the two.

interesting: $S \otimes T = U \otimes$(general coordinate transformation). Thus, we have stepped out and achieved a degree of freedom in which arbitrary deformations of space-time can be achieved: no longer do we rotate the entire Earth about one axis, but we twist and wrinkle its surface. In three dimensions, local supersymmetry is a mountain builder! Apparently, local supersymmetry includes a prescription for the geometry of space-time. This is very important because we have seen that the geometry of space-time produces what we call gravity in the large-scale world! This is why local supersymmetry is also known as supergravity.

Not only is gravity included in a supergravity theory, but we also find that other forces can be caught in the same formalism. Just as gravity is associated with the warping of space-time, so are other forces linked to the geometry of other dimensions of the vacuum. In this view, the vacuum is a fantastically complicated landscape that is symmetric under one huge local symmetry. Present indications are that there is one almost unique combination of dimension and symmetry: a ten-dimensional vacuum for fermions, 26 dimensions for bosons and symmetry groups that are either a product of two exceptional-eight groups, $E_8 \otimes E_8$, or the group known as SO(32). By the way, there is no reason *per se* to feel uneasy about a high-dimensional vacuum: it is comforting that at least the dimensions prescribed by theory do not amount to *fewer* than four, which, *a priori*, might well have been the case.

We may think of the vacuum as a multidimensional foam, the structure of which is governed by a local symmetry. The fact that geometrical curvature and symmetry are related can be demonstrated, again, by dragging our long-suffering home planet around. Earth is very nearly a sphere: it has a constant positive curvature everywhere (Fig. 14.8). This is related to a rotational symmetry, which becomes apparent when we take a patch of surface on the equator and move it straight to the North Pole, then to another point on the equator and finally along the equator back to where it came from. Then we see that the patch has rotated: repeated translations on a curved surface produce a rotation. Of course, the corresponding behaviour of a ten-dimensional vacuum is much more unruly.

How can we connect this vacuum foam structure to our world of particles? Consider the propagation of an electron. In the theories presented so far, we have always assumed that the paths are lines in space-time and that the vertices are points at a fixed time and location. But if the vacuum is really the kind of multidimensional foam we glimpsed above, we may suppose that the strands of the Feynman web are thin tubes (Fig. 14.9), in which the extra dimensions are rolled up tightly in all the vacuum dimensions except the four of space-time (then again, maybe these are rolled up, too, but on a cosmological scale).

If you scan this piece of vacuum foam with your Feynman diagram scanner, you see that quantum pairs are still being created and quanta are

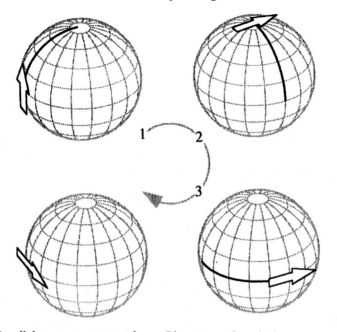

Fig. 14.8 Parallel transport on a sphere. Given a north-pointing arrow on the equator at zero longitude, we carry it around, making sure that during any displacement the new position is always parallel to the old one. First we carry the arrow to the North Pole (1 → 2), then south to the equator at 90° east (2 → 3). Then we move it back to the point of departure. Upon arrival we see that the arrow has turned 90°, even though we have always kept it parallel to itself when moving. This turning is due to the curvature of the sphere. By means of this type of process we can determine the local curvature of any surface.

exchanged, so that the picture of forces as being due to quantum exchange is still valid. However, closer inspection shows that the cross-section of a Feynman path with the scanner slot is no longer a point, but has a finite size, much like an infinitesimal cut of carrot.

Thus, in the vacuum foam picture a quantum has a small but finite extent; it extends into the extra dimensions of the vacuum. It is curled up in these other dimensions, forming a little multidimensional loop called a `superstring`.

Notice how rich the legacy is that Descartes left us, when he remarked that because space has extent, it also has substance! I wonder, though, if he would have recognized the vacuum foam as a distant descendant of his 'subtle matter'. Surely it looks very different because of the need to hide all vacuum degrees of freedom except those of space-time.

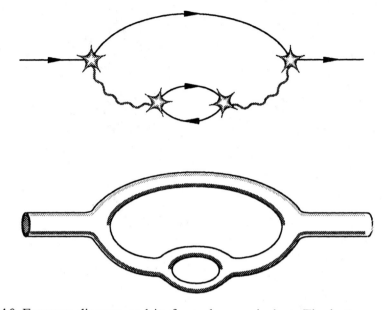

Fig. 14.9 Feynman diagram and its foamed-up equivalent. The bottom version of this interaction (a superstring diagram) does not have point-vertices. Instead, each vertex is a Y-shaped section of tubing.

Only four of the dimensions of the vacuum, the famous 'Gang of Four' of space-time, are observable on a large scale, so the others must be rolled up to a very small size, smaller than the Compton length of the known particles – maybe as small as the Planck length. The need for this so-called compactification of the higher dimensions of the vacuum may provide the explanation for particle masses. Theories of this type, in which the vacuum has more than four dimensions, are collectively known as Kaluza-Klein theories, and they work roughly as follows.

You cannot string your guitar with a sheet of plastic, but if you roll it up along one of its two dimensions you obtain a compact cylinder that is suitable as a string (Fig. 14.10). A string has a fixed number of ways in which it can vibrate and each of these has a characteristic frequency. This property makes string instruments useful for musicians because the frequency of a vibration determines the pitch of its musical note. In the quantum domain, such vibrational modes also exist; because of the quantum connection $E = 2\pi\hbar f$ between energy and frequency, and because of the relativistic $E = mc^2$, each oscillation frequency of the string represents a mass $m = 2\pi\hbar f/c^2$. All dimensions of the vacuum, except the four of space-time, must be compactified. Thus, each space-time point is associated with an

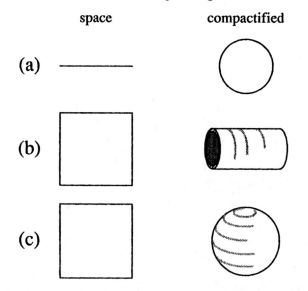

space   compactified

(a)

(b)

(c)

Fig. 14.10 Examples of compactified spaces of increasing dimension. (a) A one-dimensional space compactified into a circle. (b) A two-dimensional space compactified in one of its dimensions, producing a tube or string. (c) A two-dimensional space compactified in both its dimensions, producing a sphere.

extremely narrow supertube into which the other dimensions of the vacuum have been rolled up. There is a song from the 1940s, that goes something like the following:

He attracted some attention / When he found the fourth dimension / But he ain't got rhythm / So no one's with him / The loneliest man in town...

Such are the perils of physics; the person who wants *ten* dimensions is not likely to be treated much better!

It is surely fascinating that there might be a theory of everything some-where behind all this, but what good does it do at present? Actually, for the moment it seems that superstring theory is good-looking but useless. Certainly, it has some notable advantages. Chief among these is the fact that the particle paths split smoothly (Fig. 14.11), rather than being exact points as in the quantum field theories we discussed before. As it happens, this smoothness removes many (in some cases, all) of the renormalization infinities that plague other theories. The absence of point-vertices leads to miraculous cancellation of infinities because a smooth vertex always has a finite size, and we recall that some of the infinity problems arose because the vertices on closed loops in Feynman diagrams can be brought arbitrarily

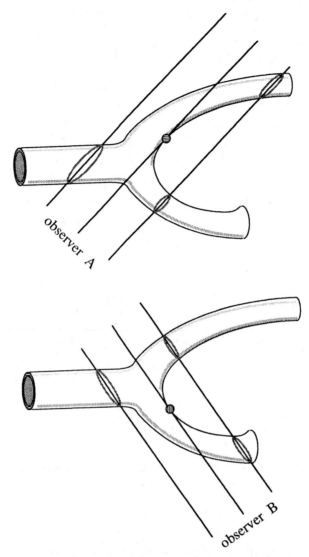

Fig. 14.11 Vertex seen from two different Lorentz frames. The particle paths split smoothly, rather than being exact points as in other quantum field theories. Thus, the vertices look the same from both Lorentz frames and are more easy to handle.

close together. Furthermore, a smooth vertex looks smooth to all observers. That is to say, no matter in what orientation the slit of our Feynman diagram scanner traverses the string splitting, we always see a smooth branching.

You may get an intuitive impression of the kind of simplification that string theories may contain, by considering simple electron–electron scattering.

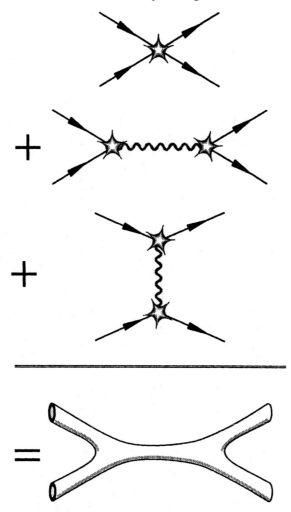

Fig. 14.12 Three Feynman diagrams of the interaction of two fermions can be replaced by a single string diagram. The double-Y tube at the bottom can be smoothly transformed into any of the three shapes drawn above it.

The three lowest-order Feynman diagrams in the classical theory can all be replaced by a *single* string diagram (Fig. 14.12). This means that the possible nasty properties of the various individual diagrams stand a much better chance of cancelling than if one were to keep the diagrams separate. This is even more dramatically evident in the case of higher-order terms: in Fig. 14.13 we see that one string diagram can stand for *nine* classical Feynman diagrams!

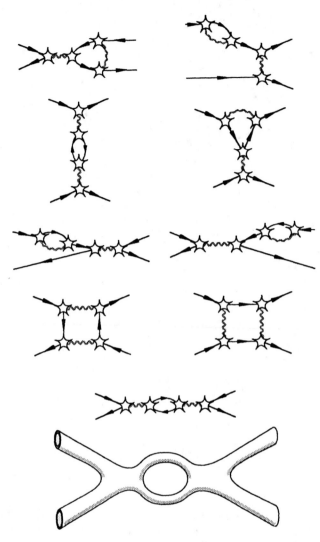

Fig. 14.13 These nine higher-order Feynman diagrams of the interaction between two fermions can be replaced by a single string diagram that contains a hole. The tube shape can be smoothly transformed into any of the nine Feynman shapes.

On the other hand, superstring theories have some notable disadvantages. They are technically extremely complicated and most of the relevant mathematics is ill developed and poorly understood. Even worse is the fact that there is not a shred of observational evidence on these theories, either for or against. The reason is that superstrings are compactified on the Planck scale of about 10^{-35} metres, with an associated energy of 10^{19} GeV. This is a phenomenal amount of energy, about equal to that of a five-kilogram meteoroid hitting Earth.

The Planck length is about 10^{22} times smaller than the Compton length of the electron: in order to see superstringiness in an electron, we would need a microscope magnifying the particle ten thousand million million million times. If we apply the same magnification to our daily world, calculating the structure of a single electron from the behaviour of the vacuum foam is equivalent to calculating the structure of the entire Solar System, including Earth and its inhabitants, from the given properties of atoms! If we could do this, we would not even have to go to Mars to see if there is life on that planet: we could simply calculate it. Of course, the comparison of length scales may be misleading, but it seems that it is a lot easier to describe the entire human body, right down to the atomic level, than to describe an electron on the basis of superstrings.

14.6 The final sentence

Even though superstrings may not blossom into a true theory of everything, it is very probable that the majority of what I have said about quantum field theories will remain valid. Although it is possible that someone might come up with a working model of the forces of Nature that includes none of the main elements in this book (relativity, quantization and symmetry), that possibility is quite remote. There is much less revolution in the development of physics than is usually admitted, so the developments in the immediate future can be expected to be rather direct continuations of present trends.

These future developments must address the current causes for concern. The main experimental source of dismay is that quantum objects do not become simpler as we look at smaller and smaller length scales. On the contrary, the symmetries we discover become more and more complex, as if squeezing a system in space makes it bulge out in another direction (a working example of what physicists call the 'law of conservation of misery'). The main theoretical source of disgust is the blatant incompleteness of it all, the lack of restrictions, the absence of compelling reasons to use one model over all the others. Every particle mass, mixing angle or special-pleading argument that must be imposed from the outside is an indictment of the theory. Why do quarks have different flavours, and why do differently flavoured quarks have different masses? Why are electromagnetism and the weak interactions mixed in the way they are? Above all, why can't gravity be included? How many dimensions are there, and what symmetries do they obey? What politicking allows only the Gang of Four to escape compactification? Is one or more of these dimensions of space-time compactified on an astronomical scale? Does the Universe expand, perhaps, because its spatial dimensions become bigger at the expense of those poor compactified others? Why is the electric fine structure constant equal to 1/137.036? And so forth: the book ends, but not the story. Fiction writers worry about first as well as last sentences, but I don't have to do that: in the book of physics, there never is a final sentence.

A modest reading proposal

Do it yourself is the main motto of physics. Not only is it fun to do so, but also it means that you don't have to believe what I say: you can go and check everything. And you should. Admittedly, it is a bit difficult to build your own particle accelerator and hunt for the Higgs boson, but throughout this book I have mentioned many things which you can verify straight away.

For those things where verification is a little more difficult, such as measuring the charge of the quarks, there is 'the literature'. This refers to a cubic kilometre or so of physics reference material which has accumulated over the course of the centuries. By reading that, you can find out what has been done in the whole field.

Because there's so much of it, it is absolutely impossible (and indeed unnecessary) to check it all out. A good way to start would be to attend introductory physics lectures, but if you want to stay home I can suggest the following course of action: on the following pages I have listed the most essential literature which I have used in my work on quantum mechanics. I will sketch a possible path through that mound of learnedness. If you follow that path, or something like it, you'll deepen and extend your physics knowledge immensely.

To begin with, there is the stuff which contains (almost) no mathematics. You should begin straight away by reading Feynman's *QED*. This book appeared while I was writing mine; if it had shown up earlier, I probably wouldn't have bothered to write what I did. Thereafter, get Feynman and Weinberg's *Elementary Particles and the Laws of Physics*, Feynman's *The Character of Physical Law*, and all of the listed articles in *Scientific American*.

You will find experimental particle physics in Fritsch's *Quarks* and Weinberg's *Discovery of Subatomic Particles*, followed by Hughes's *Elementary Particles*.

After that, you may be able to get on to the somewhat more mathematical literature. Unfortunately, there is hardly any transition-type writing: either you're hit by double-barrel maths or none at all. Anyway, you may start with Dodd's *The Ideas of Particle Physics* and follow with Bell's *Speakable and Unspeakable in Quantum Mechanics*. Some visual insight may also come from Brandt and Dahmen's *Picture Book of Quantum Mechanics*.

Next, proceed to the actual textbooks. Non-relativistic quantum mechanics from Merzbacher, then the relativistic stuff in Dirac. The detailed particle physics books, in which you will find most of the equations which I omitted from this book, can be gleaned from Feynman's *Theory of Fundamental Processes*, Veltman's *Diagrammatica*, Aitchison and Hey's *Gauge Theories in Particle Physics*, Ramond's *Field Theory*, Taylor's *Gauge Theories of Weak Interactions*, Kaku's *Quantum Field Theory*, Georgi's *Lie Algebras in Particle Physics* and especially Coleman's *Aspects of Symmetry*. By the time you're this far, you are way ahead of me and won't need any of my guidance through this jungle.

There is much material which relates more indirectly to the subjects in this book. Some of the deeper problems in classical mechanics are discussed in Mach's *Science of Mechanics*. There is the stuff on (general) relativity, such as the work by Bondi, Einstein, Geroch, McCrea, Misner Thorne and Wheeler, Pauli, Taylor and Wheeler, and Wheeler. Also, for those who like to know the people who did some of this work a little better, there are the books by Pais, Gleick, Bernstein, Crease and Mann, Pagels, and the listed material from *Physics Today*.

The rest of the stuff on my list is either front-line professional literature or material of historic or social interest. Do with it as you please.

References

Abell, G.O. and Singer, B., *Science and the Paranormal*, Scribner's, New York, 1981.

Aitchison, I.J.R. and Hey, A.J.G., *Gauge Theories in Particle Physics*, Hilger, Bristol, 1989.

Appelquist, T., Chodos, A. and Freund, P.G.O., *Modern Kaluza–Klein Theories*, Addison-Wesley, Menlo Park, 1987.

Bell, J.S., *Speakable and Unspeakable in Quantum Mechanics*, Cambridge University Press, 1987.

Bernstein, J., Spontaneous symmetry breaking, gauge theories, the Higgs mechanism and all that, *Reviews of Modern Physics*, **46**, 7, 1974.

Bernstein, J., *The Tenth Dimension*, McGraw-Hill, New York, 1989.

Bondi, H., *Relativity and Common Sense*, Heinemann, London, 1964.

Brandt, S. and Dahmen, H.D., *The Picture Book of Quantum Mechanics*, Wiley, New York, 1985.

Clauser, J.F. and Shimony, A., Bell's theorem: experimental tests and implications, *Reports on Progress in Physics*, **41**, 1881, 1978.

Coleman, S., *Aspects of Symmetry*, Cambridge University Press, 1993.

Crease, R.P. and Mann, C.C., *The Second Creation*, Macmillan, New York, 1986.

Dawkins, R., *The Blind Watchmaker*, Penguin, London, 1986.

Descartes, R., *Œvres et Lettres*, Gallimard, Paris, 1953.

Devlin, K., *Mathematics: the New Golden Age*, Penguin, London, 1988.

Dirac, P.A.M., *The Principles of Quantum Mechanics*, 3rd edn, Clarendon Press, Oxford, 1956.

Dodd, J.E., *The Ideas of Particle Physics*, Cambridge University Press, 1985.

Einstein, A., *The Meaning of Relativity*, Princeton University Press, 1956.

Englert, F. and Brout, R., Broken symmetry and the mass of gauge vector mesons, *Physical Review Letters*, **13**, 321, 1964.

Feynman, R.P., *The Character of Physical Law*, M.I.T. Press, Cambridge, Massachusetts, 1965.

Feynman, R.P., *QED*, Princeton University Press, 1985.

Feynman, R.P., *The Theory of Fundamental Processes*, Benjamin/Cummings, Reading, 1962.

Feynman, R.P. and Hibbs, A.R., *Quantum Mechanics and Path Integrals*, McGraw-Hill, New York, 1965.

Feynman, R.P., Leighton, R.B. and Sands, M., *The Feynman Lectures on Physics*, Addison-Wesley, Reading, 1977.

Feynman, R.P. and Weinberg, S., *Elementary Particles and the Laws of Physics*, Cambridge University Press, 1987.

Frey, A.H. and Singmaster, D., *Handbook of Cubik Math*, Enslow Publishers, Hillside NJ, 1982.

Fritzsch, H., *Quarks*, Penguin, Harmondsworth, 1983.

Froggatt, C.D. and Nielsen, H.B., *Origin of symmetries*, World Scientific, Singapore, 1991.

Gamow, G., *Mr Tompkins in Paperback*, Cambridge University Press, 1971.

Georgi, H., A unified theory of elementary particles and forces, *Scientific American*, p.48, April 1981.

Georgi, H., *Lie Algebras in Particle Physics*, Benjamin/Cummings, Menlo Park, 1982.

Geroch, R., *General Relativity from A to B*, University of Chicago Press, 1978.

Gleick, J., *Chaos*, Penguin, London, 1987.

Gleick, J., *Genius*, Pantheon, New York, 1992.

Gould, S.J., *Ever Since Darwin*, Norton, New York, 1977.

Green, M.B., Unification of forces and particles in superstring theories, *Nature*, **314**, 409, 1985.

Greenberg, O.W., A new level of structure, *Physics Today*, **38**, 22, 1985.

Haynes, R.H. and Hanawalt, P.C. (eds.), *The Molecular Basis of Life*, Freeman, San Francisco, 1968.

Heilbron, J.L., Bohr's first theories of the atom, *Physics Today*, **38**, 28, 1985.

Hendry, A.W. and Lichtenberg, D.B., The quark model, *Reports on Progress in Physics*, **41**, 1707, 1978.

Higgs, P., Broken symmetries and the masses of gauge bosons, *Physical Review Letters*, **13**, 508, 1964.

Higgs, P., Spontaneous symmetry breakdown without massless bosons, *Physical Review* 145, 1156, 1966.

Hughes, I.S., *Elementary Particles*, Cambridge University Press, 1991.

Miller, A.I., *Early Quantum Electrodynamics*, Cambridge University Press, 1995.

Lipkin, H.J., *Lie Groups for Pedestrians*, North-Holland, Amsterdam, 1965.

McCrea, W., Time, vacuum and cosmos, *Quarterly Journal of the Royal Astronomical Society*, **27**, 137, 1986.

Mach, E., *The Science of Mechanics*, Open Court, LaSalle, 1960.

Mandl, F., *Introduction to Quantum Field Theory*, Wiley/Interscience, New York, 1959.

Mermin, N.D., Is the Moon there when nobody looks?, *Physics Today*, **38**, 38, 1985.

Merzbacher, E., *Quantum Mechanics*, Wiley, New York, 1961.

Misner, C.W., Thorne, K.S. and Wheeler, J.A., *Gravitation*, Freeman, New York, 1973.

Morrison, P., Morrison, P., Eames, C. and Eames, R., *Powers of Ten*, Scientific American Library, San Francisco 1982.

Pagels, H.R., *The Cosmic Code*, Penguin, Harmondsworth, 1982.

Pais, A., *Subtle is the Lord...*, Oxford University Press, 1982.

Pais, A., *Inward Bound*, Clarendon Press, Oxford, 1986.

Pais, A., *Niels Bohr's Times*, Clarendon Press, Oxford, 1991.

Pauli, W., *Theory of Relativity*, Dover, New York, 1958.

Ramond, P., *Field Theory*, Addison-Wesley, Redwood City, 1989.

Schutz, B.F., *Geometrical Methods of Mathematical Physics*, Cambridge University Press, 1990.

Schrödinger, E., *What Is Life?*, Cambridge University Press, 1983.

Schwarz, J.H. (ed.), *Superstrings* (two volumes), World Scientific, Singapore, 1985.

Schweber, S.S., *QED and the men who made it*, Princeton University Press, Princeton, 1994.

Taylor, J.C., *Gauge Theories of Weak Interactions*, Cambridge University Press, 1976.

Taylor, E.F. and Wheeler, J.A., *Spacetime Physics*, Freeman, New York, 1966.

't Hooft, G., Gauge theories of the forces between elementary particles, *Scientific American*, p.104, June, 1980.

Utiyama, R., Invariant theoretical interpretation of interaction, *Physical Review*, **101**, 1597, 1956.

Veltman, M., *Diagrammatica*, Cambridge University Press, 1994.

Weinberg, S., *The Discovery of Subatomic Particles*, Scientific American Library, New York, 1983.

Wheeler, J.A., *A Journey into Gravity and Spacetime*, Scientific American Library, New York, 1990.

Yang, C.N. and Mills, R.L., Conservation of isotopic spin and isotopic gauge invariance, *Physical Review*, **96**, 191, 1954.

Glossary

Mathematicians define things, otherwise they wouldn't have a clue what they are talking about. This is so because everything in mathematics was invented by people. Contrariwise, nothing of the substance of physics was invented by us: Nature is out yonder and there is not a shred of evidence that the Universe cares a fig about human or other beings. The *formulations* of physics are uniquely ours, but those, of course, are mathematics.

That is why, contrary to popular opinion, physicists never actually define anything physical. There are whole walls in libraries covered by countless shelves which are bending under the books on electronics and quantum electrodynamics. But in none of those books will you find a proper definition of an electron, for the very good reason that we haven't the foggiest idea what an electron 'is'. In fact, we do not know what such an existential question is supposed to mean, and how we would judge a particular answer to it as being right or wrong; personally, I do not think that such questions belong to physics at all.

When we talk about a physical thing such as an electron we 'define' it by an indirect approach, mostly by talking about its properties (mass, electric charge, spin,...) and how these differ from the corresponding properties of other physical things. For example, there is a particle – the muon – which, as far as we know today, is identical to the electron in every way except for the fact that it has a mass that is a few hundred times bigger. We can perform numerous experiments trying to refine that difference, linking it to other properties of other particles. We expect that eventually some pattern of similarities, differences and interrelations between them appears that shows a texture we can recognize and interpret and which guides us in making predictions. Thus we gain insight in the significance of the differences between electrons and muons, but we still don't know what they 'are'.

The second indirect way of 'defining' something in physics is by giving a description of what happens under certain specified circumstances. That is why physicists describe in minute detail what they did and why they attach great value to the ability to repeat experiments. In this manner a particle is 'defined' by describing its interactions with other particles: *electron is as electron does.*

It is obvious that these necessarily indirect 'definitions' or circumscriptions imply that we must be familiar with the majority of contemporary particle physics before we can get a grasp of what these strange beasts called electrons and their ilk 'are'. This requires the kind of assiduous study that is not for the feebly motivated or the fainthearted. It follows that at some level the casually interested person must be content to accept certain things on the 'authority' of those who have studied physics more closely. It also means that if you continue to ask probing questions there *always* comes a point when the interrogated physicist will say (at least if s/he is honest): 'I haven't the foggiest idea.' To the professional, that's where the real fun begins!

Hence the pained and perplexed look on the face of a physicist when confronted with a question such as 'But what *is* space, really?' In response I can begin by explaining that we don't actually define things (this causes disbelief and consternation); continue by enumerating some of the things that lead us to believe that there is such a thing as space, for example the cheerful fact that I do not stand on everybody else's toes all the time; point out that the observed fact that one point in space is as good as any other implies the existence of funny things such as momentum and spin which have properties that can be verified with great accuracy (this causes acute distress); and, finally, that the only way to get some working knowledge of what space 'is' would be to go study physics for a couple of decades (this occasionally causes outright hostility).

Accordingly, the following pages are not meant as a list of definitions. It is best to regard them as an extremely brief synopsis of some of the things said in the book, with an indication of the pages on which they are introduced or expanded upon. Of course, for all the reasons listed above, there is a lot of vagueness, indirectness and circularity because one cannot fairly expect to find the whole canon of physics in one book. Thus, this list is mostly meant as an *aide-mémoire* and as a small road map for the book, arranged by subject instead of sequentially like the table of contents.

```
       In the book, the entries below
     are identified by this typeface.
```

`Abelian` **[105,111]** If two mathematical operations, performed sequentially, produce the same result independent of the order in which they are executed, they are said to commute. A group of which the multiplication operation obeys this restriction is commutative or Abelian: $a \otimes b = b \otimes a$ (see `symmetry group`).

`acceleration` **[47]** The rate at which the velocity changes with time: the *first time derivative* of the velocity. Observe a velocity, do the same an instant later, subtract the velocities (by means of vector subtraction) and divide the difference by the time difference between the two positions. Thus, the acceleration is a vector with the same number of components as the velocity. Because of its definition, the acceleration is the *second time derivative* of the position.

`action` **[50,91]** In classical mechanics, the kinetic energy minus the potential energy of a particle, summed (integrated) over the time of flight along the particle's path. The path a particle takes, out of all possible paths, is the one along which the action is least. In the Feynman formulation of relativistic quantum mechanics the action is the integral over time of the 'action density' of a particular interaction along a path. The action density is a mathematical expression which summarizes all the ways in which a given set of quanta can interact. In a theory based on local symmetry, the form of the action density – if one indeed exists – is prescribed (sometimes exactly) by the requirements of relativity, quantization and local symmetry. Using the quantum equivalent of the principle of least action one can derive the equations of motion of the fields (see `Feynman path`).

`alternatives` **[55]** There is an absolute scale of smallness in Nature: small things are necessarily perturbed by an interaction. This ineradicable influence can also be expressed as an uncertainty relation. Uncertainty implies that any small-scale process can occur via several alternatives.

`amplitude` **[32]** (of a wave) Half the height between the crest of a wave and the adjacent trough.

`amplitude` **[57]** (of a quantum process) Every alternative of a process can be assigned a complex or 'imaginary' number a that is related to the probability P of that alternative by the relation $P = a^* a$ (the index $*$ denotes an operation called 'complex conjugation'). The number a is the amplitude of the alternative. The linear sum of all its alternatives gives the total amplitude of a process (see `complex numbers`).

`angular momentum` **[106,107]** A vector quantity, often denoted as $\vec{J}$: the product of the mass m, the velocity $\vec{v}$ and the position $\vec{r}$ of a particle. The position can be measured from any point; thus, the value of $\vec{J}$ depends on where that point is. The direction of $\vec{J}$ is perpendicular to both $\vec{v}$ and $\vec{r}$, and forms a right-handed triple with these. Let all three vectors start at the centre of the Earth; if $\vec{v}$ points to the intersection of the equator and the null meridian, and if $\vec{r}$ points to the North pole, then $\vec{J}$ points to the equator at 90° west. The total angular momentum of a closed system is constant (conserved). The existence and conservation of angular momentum is a consequence of Noether's theorem applied to the rotational symmetry of space.

`annihilation` **[124]** (see `antimatter`).

`antimatter` **[121]** Due to relativity the time order of events is not absolute but relative: if event L happens before H as seen by some observers, others may see H occur before L. Due to quantum uncertainty, this may even happen for points on the path of a particle that moves more slowly than light. Thus, there is a certain probability that a particle's path is seen to be 'doubled over' as a Z-shape in the space-time of some observers. These then see, at the first corner of the Z, two objects appear simultaneously: this is called pair creation. One of these is arbitrarily labelled a particle, the other is called its antiparticle. At the second point of the Z, two particles disappear simultaneously; this is called annihilation.

`associative` **[104]** If three instances of a mathematical operation, performed sequentially, produce the same result independently of the way in which they are grouped, the operation is associative: $(a \otimes b) \otimes c = a \otimes (b \otimes c)$ (see `symmetry group`).

`asymptotic freedom` **[265]** The observation that the colour force becomes weaker and weaker at distances smaller than those between quarks in a baryon.

`bare charge` **[256]** (naked charge) The electric charge used in the amplitude of a particle between vertices in a single Feynman diagram. This differs from the charge observed on large scales, because the process we observe is the quantum sum of all its possible Feynman alternatives. The effective charge thus observed at large is the dressed charge. The dressed charge is finite; the bare charge may, in the limit for very small-scale processes, become infinite.

`bare mass` **[252]** The mass used in the amplitude of a particle between vertices in a single Feynman diagram. This differs from the mass observed on large scales, because the process we observe is the quantum sum of all its possible Feynman alternatives. The effective mass observed at large is the dressed mass, which is finite; the bare mass may, in the limit for very small-scale processes, become infinite. Via the equivalence of mass and energy, this infinite mass is seen to correspond to the infinite self-energy of the particle.

`baryon` **[28,210]** A composite of three quarks, held together by the colour force.

`beta decay` **[220]** The name derives from a certain type of radioactivity, in which an electron e^- (beta particle) is emitted by a nucleus. On a particle scale, beta decay is due to a weak-force process in which a d-quark changes flavour and becomes a u-quark, as in $(udd) \rightarrow (uud) + e^- + \nu$ (neutron decay).

`binding energy` **[228]** The energetic difference between two ways of arranging several particles. Because of the relativistic equivalence of mass and energy the binding energy shows up at large as a difference in mass. For example, the mass of two neutrons plus two protons is larger than the mass of a helium nucleus, which consists of those particles bound together by the strong nuclear force. This does not mean that the individual particles in a bound state have different masses; the total mass is a property of the aggregate as a whole.

`black hole` **[204]** A space-time structure in which the light cones, as seen from a point infinitely far away, are tilted so strongly that they enclose a region of space (called a 'closed trapped surface' or 'horizon') from within which light cannot escape to infinity. Because nothing can go faster than light, all particle trajectories must stay within the light cone so that nothing can escape from the region enclosed by such a horizon, whence the expression 'black' hole.

`Bose-Einstein particle` **[140]** (see `boson`).

`boson` **[19,140]** A particle with integral spin: $S = 0, 1, 2, \cdots$. These obey 'Bose–Einstein statistics': given N identical particles in a quantum state, the probability that yet another one will find a place in that state is multiplied by a factor $N + 1$. Thus, bosons are 'gregarious' particles.

`Cabibbo rotation` **[245]** Weak processes may change the flavour of quarks. Thus, we must think of a quark as a superposition of several flavours. Therefore, a transition such as $d \rightarrow s$ can take place spontaneously; this is a Cabibbo rotation (called rotation because of the occurrence of a mixing angle in the superposition).

`charge` **[190]** A measure of the strength of the coupling between particles. The total charge of a closed system is constant (conserved). The existence and conservation of charge is a consequence of Noether's theorem applied to an internal symmetry of particles.

`charge renormalization` **[256]** The process whereby the electric charge of virtual particles is adjusted in such a way that the charge observed at infinity remains finite (see `bare charge`).

charm quark **[247]** (see flavour).

chemical affinities **[161]** A superposition of electrons in which all angular momentum states are filled is especially stable. Atoms that can provide the electrons for such combinations are more likely to bind together than others. The extent to which atoms engage in this binding is called affinity.

chimaera **[94]** A beast in Greek mythology with the head of a lion, the body of a goat and the tail of a serpent. Used as a metaphor for quantum superposition.

colour charge **[208]** A measure of the strength of the 'strong' coupling between particles; a consequence of Noether's theorem applied to the internal SU(3) symmetry of particles (see charge).

commutative **[105,111]** (see symmetry group).

compactification **[288]** There are more things than their location that distinguish particles from one another; according to some speculative modern theories, these things are the expression of a richer structure of space than mere extent. The fact that in our large-scale world we cannot travel in these extra spatial directions means that they must be rolled up or compactified into very small structures.

complementary **[51,62]** The exact observation of certain quantities excludes the simultaneous observation of certain others. Bohr called a pair of such quantities complementary. In the literature they are also called conjugate quantities. The variables corresponding to these in mathematical equations are conjugate operators.

complex number **[65,82]** A two-dimensional number obeying certain mathematical rules and regularities. A complex number c can be related to the ordinary 'real' number continuum by $c = p + iq$, in which p and q are real numbers (such as π or 2 or 2.718) and where the item i has the property $i \times i = -1$. Complex numbers c and $d = r + is$ can then be added as $c+d = (p+r)+i(q+s)$ and multiplied as $c \times d = (p \times r - q \times s) + i(p \times s + q \times r)$. The operation called 'complex conjugation', indicated by an asterisk, consists of inverting the sign of the factor containing i: if $c = p + iq$, then $c^* = p - iq$. The absolute value A of a complex number c is then easily found as $A^2 = c^* \times c = p^2 + q^2$.

complex probability amplitude **[82]** (see amplitude).

components **[12]** The numbers representing the strength of a field. A scalar field (such as the temperature) consists of a single component. Vector fields (such as wind velocity) consist of a row of components (three for a wind field in three-dimensional space). Tensor fields (such as the shear in Earth's crust when it is deformed by geological processes) consist of a rectangular array (matrix) of components.

Compton length [125] Relativistic counterpart of the De Broglie length. Usually indicated by λ_C, it can be calculated from the speed of light c and the mass m of a particle by $\lambda_C = \hbar/mc$ (occasionally used as $\lambda_C = 2\pi\hbar/mc$). Because the speed of light cannot be exceeded, the Compton length is the 'absolute size' of a particle (see De Broglie length).

conjugate variables (operators) [62] (see complementary).

conserved quantity [105] Quantity that remains the same in a closed system (see Noether's theorem).

coordinates [46,114] Numbers that specify the position of a point. They are usually chosen to conform to a known symmetry of a particular situation, such as spherical or cylindrical coordinates. Coordinates that form a regular rectangular network are called Cartesian (see space).

Coulomb gauge [262] A particular choice for the representation of a virtual photon, in which the photon is linearly polarized perpendicular to its direction of motion. This is an example of a 'fixed gauge'.

Coulomb's law [258] The law of static attraction or repulsion between two electric charges a distance R apart: (force)= constant/R^2.

coupling constant [191] A measure of the strength of the coupling between particles (see charge).

covalent binding [163] A type of chemical attraction due to the fact that the simultaneous (partial) presence of two electrons in a certain region of space can be energetically advantageous. It is the weakest type of chemical bond. Because the electron overlap required in covalent binding demands close contact between molecules, covalent chemical reactions depend very sensitively on the shape of the participating molecules. Thus, these reactions are essential for the specificity of reactions in living beings.

current [261] The amount of charge that flows in a unit of time through a unit of spatial surface.

curvature of space-time [200] A scalar quantity that is a measure of the structure of space and time. The full structure is expressed by a tensor from which the curvature can be derived (see metric tensor).

De Broglie length [55] A length scale below which things are properly called small. Usually indicated by λ, it can be calculated from the speed v and the mass m of a particle by $\lambda = h/mv = 2\pi\hbar/mv$. (see Compton length).

decomposition **[94]** In quantum mechanics, a state can always be thought of as a superposition of other states. The prescription for such a superposition is a decomposition of that state in terms of others: how is a raven like a writing-desk?

degenerate **[235]** When two states with different quantum numbers have the same energy they are said to be degenerate with each other. For example, in the classical model of the electron in a hydrogen atom the energy of the states depends only on the principal quantum number *n*, not on *l* or *m*. Degeneracy is the signature of a high degree of symmetry (spherical symmetry in the case of hydrogen). Subtle effects (such as interaction between the spin of the electron and the spin of the nucleus) can break that symmetry and lift the degeneracy. This leads to a small energy difference between the formerly degenerate states. Because of the equivalence of mass and energy this is analogous to giving mass to particles by spontaneous symmetry breaking.

deoxyribonucleic acid (DNA) **[171]** Twisted pair of long chains (polymer) of molecular building blocks, cross-linked by other molecules called bases. The sequence of the bases is a code for the construction of other biological molecules. The building blocks in each chain are pentose sugar molecules, linked lengthwise by phosphate groups to form a very long strand. The bases, which link two opposite-running sugar–phosphate strands into a helical ladder, are four possible molecules: adenine, cytosine, guanine and thymine.

diffraction **[36]** If a wave is partly blocked by an obstacle, the wave curls around it. This is one instance of diffraction. By arranging the obstacles in special ways (in particular by making the arrangement regular, such as in a grating or a hologram) the diffraction can be made to produce special effects by interference.

diffraction grating **[23]** A regular array of narrow openings that uses diffraction to send waves with different wavelengths in different directions. Used to sort waves by wavelength.

dimension **[177]** (see symmetry group).

dimensions **[46]** Numbers required to specify the position of a point in space. In our Universe, space requires three numbers: space is three-dimensional. Space-time requires one more and is four-dimensional. According to some speculative modern theories, space-time has a richer structure than mere extent. The corresponding vacuum would then be a multidimensional space (as many as 26 dimensions have been proposed).

dipole field **[217]** Field obtained by placing a positive and an equally sized but negative charge closely together.

dispersion **[67]** The fact that the propagation speed of a wave depends on its wavelength or frequency. This effect allows a prism to sort light by wavelength.

distance **[46]** Separation of points in space; in space-time the distance is called interval.

doublet **[105,175]** (see multiplet).

down quark **[212]** (see flavour).

down spin **[146]** (see spin).

dressed charge **[257]** (see bare charge).

dressed electron **[252]** An electron as seen by an observer on large scales. Because any process is the quantum sum of all its Feynman alternatives, the dressed electron differs in some respects from the 'bare electron', which is the particle which occurs in each Feynman diagram individually.

dressed mass **[252]** (see bare mass).

eigenstate **[94]** Given a quantum state called L. Suppose that a certain measurement, performed on L, yields with absolute certainty the quantity l as outcome. Given a state G; suppose that the same measurement, performed on G, yields with certainty the quantity g. Then L and G are eigenstates ('pure' states) of that particular measurement, with eigenvalues l and g.

eigenvalue **[94]** (see eigenstate).

electric charge **[190]** A consequence of Noether's theorem applied to the phase rotation symmetry U(1) of particles (see charge).

electron **[28]** One of the main constituents of ordinary matter. It is a fermion with spin-$\frac{1}{2}$, mass 0.5 MeV/c^2 and electric charge -1. At this moment the electron is not known to have any substructure and can be called a true 'elementary' particle.

electron shells **[159]** Electrons in atoms must occupy specially shaped regions prescribed by the rotational symmetry of space. Each shape corresponds to a particular energy. Because no two electrons can have the same quantum numbers, additional electrons must occupy higher and higher energy states, thus forming electron shells (see Pauli exclusion principle).

electron volt **[165]** Energy acquired by an electron when passing through a potential drop of one volt. It is 1.6×10^{-19} joule, so that one exa-eV, a billion billion eV (1 EeV $= 10^{18}$ eV), corresponds to this book dropped from a height of about ten centimetres. Using the mass–energy equivalence $E = mc^2$ we can express a mass in eV (usually GeV $= 10^9$ eV) by multiplying by c^2. The mass associated with one GeV is then 1.78×10^{-27} kg. The mass of the proton is 0.938 GeV; the mass of the electron is 511 keV.

electrostatic force [263] (see Coulomb's law).

electroweak force [238] Unification of electricity, magnetism and the weak nuclear force. The basic multiplet is supposed to be a doublet; the corresponding velpons which transmit the force are derived from the symmetry product group U(1)⊗SU(2). Through judicious superposition (by means of the mixing angle θ_W) of the basic velpons one obtains a massive triplet of vector bosons W^+, W^-, Z and a massless photon γ.

elements [103] A symmetry group is a set of objects, called elements, plus some prescription for relating these elements to each other (see symmetry group).

energy (conservation) [50] In classical mechanics, half the product of the mass and the square of the velocity of a particle. The total energy of a closed system is constant (conserved). The existence and conservation of energy is a consequence of Noether's theorem applied to the translational symmetry of time: it doesn't matter where you choose the zero point of time (see translational invariance).

energy desert [283] The range of energies between the mass of the weak-interaction velpons (around 100 proton masses) and the hypothetical X-boson at 10^{15} proton masses.

equal-time spaces [206] Sections of space-time obtained by cutting a space-time structure by a plane at fixed time.

equation of motion [48] Relationship between the acceleration experienced by a particle and the force acting on it.

excited state [153] (see state).

Fermi–Dirac particle [141] (see fermion).

fermion [19,141] A particle with half-integral spin: $S = \frac{1}{2}, \frac{3}{2}, \frac{5}{2}, \cdots$. These obey 'Fermi–Dirac statistics': given one particle in a quantum state, the probability that yet another identical one will adopt that state is zero. Thus, fermions are 'intolerant' particles.

Feynman diagram [15] Symbolic graphical summary of one particular alternative of a given quantum interaction. Ingoing states are connected to outgoing ones by lines that represent the propagation of fermions and bosons; these are coupled at points called vertices. To each element of a Feynman diagram corresponds a well-prescribed mathematical quantity that represents the amplitude of that element; the amplitude of the entire diagram is obtained by multiplication of these quantities. The total amplitude of the interaction is then found by summing over the inifinity of Feynman diagrams that have the same in- and out-states.

Feynman path **[68,91]** Line through space-time (world line) that represents one particular alternative of the trajectory of a particle between two fixed points A and B. Each path is characterized by the value along that path of a scalar energy function, the action S. The phase ϕ of the particle is determined by $\phi = S/\hbar$ so that the amplitude for that path becomes $\exp(i\phi)$. The total amplitude for the particle's propagation from A to B is then found by linear superposition of all path amplitudes.

field **[11]** A kind of 'messenger substance' that carries the information about the action of a force from one point to another. In mathematical representation, a list of numbers that describe, at each point, the action of the force. In quantum theory, a mathematical operator representing a particular type of particle such as an electron or a photon.

field density **[92]** A quantity that, when multiplied by a small step along a path in spacetime and divided by Planck's constant, yields the change in the phase of a particle travelling along that path.

fine structure constant **[273]** (electromagnetic) The quantity $\alpha = e^2/\hbar c$, where e is the unit of electric charge. It is a dimensionless quantity, i.e. a pure mathematical number such as π or $\sqrt{2}$. The number α plays an important role in the description of all electromagnetic phenomena.

fixing a gauge **[145,147,224]** Picking one possible orientation of coordinates for doing definite calculations in a symmetric system. For example, Earth is a sphere, but we can only describe the locations on its surface if we pick a definite coordinate net, such as the usual system of latitudes and longitudes. Fixing a gauge singles out certain directions which seem to be 'special', but their specialty is due to the gauge fixing and not to an intrinsic property. In the case of latitude and longitude on Earth, the North and South poles seem special, even though on a sphere no point is different from another (see gauge group).

flat space-time **[200]** Space-time in which the light cones, viewed from infinity, all point in the same direction.

flavour **[200]** For reasons unknown, Nature consists of a triple set of 'fundamental' particles. In the case of quarks, the SU(2) doublet (u,d) (also called up, down) is tripled by the presence of (s,c) (strange, charm) and (t,b) (truth, beauty) or (top, bottom). Collectively, these six types are called flavours.

`free-fall line` **[210]** The symmetry axis of the local light cone. A particle moving along this line in curved space-time experiences no local acceleration.

`frequency` **[32]** Number of oscillations of a wave in one second.

`fringes` **[39]** Alternating ridges of high and low amplitude seen when waves interfere.

`fundamental` **[28]** Standing wave with the longest wavelength or lowest frequency in a given resonance situation.

`gauge boson` **[183]** (see `gauge field`).

`gauge field` **[183]** Field corresponding to a gauge symmetry. The symmetry is described by a gauge group. The compact gauge groups known to occur in Nature are U(1), SU(2) and SU(3); the non-compact gauge group is the Lorentz group. The fields generated by local symmetries of these types are the photon, the *W*- and *Z*-bosons, the gluons and the gravitational field respectively. A physical system that remains the same under a gauge transformation is said to show gauge invariance. For example, a closed system is invariant under the phase-changing symmetry U(1) (see `symmetry group`).

`gauge group` **[183]** (see `gauge field`).

`gauge invariance` **[147]** (see `gauge field`).

`general theory of relativity` (GRT) **[202]** Gauge field theory based on local symmetry under the Lorentz group. The GRT describes the structure of space-time by means of a distance recipe: a metric of distances in a small patch of space-time, chosen in such a way that locally the speed of light remains the same (local Lorentz invariance). The GRT also specifies how the presence of matter influences the space-time around.

`gluino` **[280]** Supersymmetric partner of the gluon.

`gluon` **[209]** Velpon that arises because of the local SU(3) symmetry of Nature. The dimension of the symmetry implies the existence of three charges, called colour charges R, G, B, and the existence of $3^2 - 1 = 8$ different gluons. Exchange of gluons mediates a force called the colour force.

`Goldstone boson` **[232]** Massless spin-1 particle that must occur in any locally symmetric gauge theory if the symmetry is unbroken.

`grand unified theory` (GUT) **[273]** Local gauge theory that attempts to encompass all known quantum interactions: electromagnetism, weak force and colour (strong) force.

gravitational constant **[205]** Coupling strength G of the gravitational field. In Newtonian mechanics, the gravitational force F between two point masses m and M a distance R apart is $F = GmM/R^2$. In general relativity, G is a proportionality factor to be multiplied by the mass density in order to obtain an expression for the space-time curvature generated by that mass.

graviton **[205]** Hypothetical particle that mediates the gravitational force. No quantum theory of gravity exists, so the properties of the graviton are uncertain. Most likely it is a tensor particle with spin 2.

gravity **[198,201]** Historical name for the immediately observable consequences of the curvature of space-time.

ground state **[153]** (see state).

group **[103,104]** (see symmetry group).

group multiplication **[103]** With every pair of elements of a group is associated a third one. The association is known as group multiplication, or multiplication for short, indicated by $\otimes$. From two elements a and b we can form $a \otimes b = c$ (see symmetry group).

gyroscope **[129]** A carefully crafted top, usually mounted in gimbal rings, used to demonstrate the peculiar properties of spinning objects.

hadron **[207]** Bound aggregate of quarks. The two known families are mesons (made of two quarks) and baryons (three quarks).

handedness **[145]** (see spin).

harmonics **[28]** Standing waves other than the fundamental which occur in a given resonance situation.

Heisenberg's uncertainty relations **[61]** (see uncertainty relations).

helicity **[145]** (see spin).

Higgs boson **[236]** Scalar particle belonging to a fundamental family symmetric under SU(2)⊗U(1) (see Higgs mechanism).

Higgs mechanism **[234]** (Englert–Brout mechanism) The hypothesis that there is a fundamental family of spin-0 particles which are symmetric under SU(2)⊗U(1), which serve to break the group symmetry. The breaking of the symmetry shows up in that three of the velpons of the SU(2)⊗U(1) field, the W^{+-}- and Z-bosons, acquire mass. The fourth, the photon γ, remains massless, so that we expect there to be a loose Higgs boson around somewhere. Up to 1994 it has not yet been found.

horizon **[204]** Surface of a region in space-time from within which nothing, not even light, can escape (see black hole).

`Huygens's principle` **[34]** The propagation of a wave can be calculated by postulating that each point of a wave crest is a source of a spherical wavelet with the same frequency and phase. By taking the envelope of all these wavelets, the position of the next wave crest can be determined.

`indeterminate` **[72,75]** The uncertainty relations show that there are pairs of physical quantities, called complementary or conjugate pairs, with the property that the precise knowledge of one of them precludes all knowledge of the other. If one property is known, its conjugate is indeterminate (to be distinguished from merely 'unknown').

`inflationary transition` **[226]** Hypothetical stage in the very early evolution of our Universe. It is supposed that the differentiation of the originally symmetric Universe into one that had separate – and partly broken – symmetries such as U(1), SU(2) and SU(3) caused the creation of massive particles. The energy corresponding to this particle creation caused the Universe to go through a very brief phase of extremely rapid expansion.

`interference` **[39,56]** When two waves are added by linear superposition, in some places wave crests coincide with other crests, leading to an increase in the amplitude, while in other places crests coincide with troughs and produce small or even zero amplitude. This is called interference. Also said of the superposition of the amplitudes of quantum alternatives.

`interference term` **[74]** Algebraic quantity that indicates the extent to which quantum alternatives interfere. For example, if a and b are quantum amplitudes, the outcome of an experiment to determine the former gives the probability a^2, while determining the latter yields b^2. If both alternatives are possible simultaneously, the probability is $(a+b)^2 = a^2 + b^2 + ab + ba$. The term $ab + ba$ is the extent of the interference between a and b.

`invariant` **[105]** Something that remains unchanged under the action of a group. For example, distances are invariant under rotations (otherwise a map of our rotating Earth would look very peculiar). Conversely, the discovery of an invariant leads to the suspicion of the existence of a corresponding symmetry group; for example, the invariance of the speed of light produces the Lorentz symmetry of space-time.

`inverse` **[104]** Each group element a has an inverse (usually called a^{-1}) such that $a \otimes a^{-1} = e$ (see `symmetry group`).

`inverse beta decay` **[220]** Particle process in which an electron (beta particle) is absorbed, such as the formation of a neutron out of a proton and an electron (see `beta decay`).

`ionic binding` **[163]** Chemical attraction between atoms in which one has acquired an electron (or more) of the other, so that one atom becomes positively charged and the other negative. The resulting electrostatic attraction binds the atoms together. It is the strongest chemical binding.

isospin (symmetry group) [177,192] The special unitary group of dimension two, SU(2), is also called isospin group. The particle property that is modified by the action of this group is called isospin ('isotopic spin'). For example, the proton and the neutron (which have very nearly the same mass and act quite similar under the strong nuclear force) are an isospin doublet in which the proton is labelled 'isospin up' and the neutron 'isospin down'. The action of SU(2) causes an isospin rotation (analogous to a phase change) that can turn the proton, via intermediate superpositions, into a neutron and vice versa.

isospin rotation [192] (see isospin).

jet [267] Closely spaced bunch of hadron tracks with a shape reminiscent of a witch's broom. These sprays are the decay products of fundamental particles in a colour interaction which cannot occur freely on large scales, such as gluons or quarks.

K-capture [157,220] Nuclear process in which the nucleus absorbs an electron from the innermost electron shell of an atom.

K-shell [220] Innermost electron shell in an atom. Electrons in the K-shell are the most tightly bound of all atomic electrons.

Kaluza–Klein theory [288] Theory of elementary forces based on the hypothesis that the vacuum has more dimensions than the four of space-time.

Kepler's second law [101] If one connects the Sun with a planet moving in its elliptical orbit, the connecting line sweeps out equal surfaces of the orbital plane in equal times due to the conservation of angular momentum.

kinetic energy [49] Energy of motion. In classical mechanics, the kinetic energy E_k of a particle with mass m and speed v is given by $E_k = \frac{1}{2}mv^2$.

left-handed [145] (see spin).

leptons [28] Low-mass particles such as the electron, the muon, the tau and their corresponding neutrinos, that do not feel the colour force.

light cone [119,198] Surface in space-time along which light propagates. The mathematical equation for the light cone surface is $x^2+y^2+z^2-c^2t^2 = 0$.

linear waves [37] In classical mechanics, waves with such small amplitudes that they can be superposed linearly, i.e. by direct addition of their amplitudes. In quantum mechanics all quantum waves are linear, because the uncertainty relation requires that their superposition must not carry any traces of the order in which the superposition was done.

longitudinal photon **[262]** Photon polarized with its electric component oscillating in its direction of motion. Relativity shows that, because a photon has no rest mass, a real (free) photon cannot have longitudinal polarization. However, the uncertainty relation allows a virtual photon to have a longitudinal degree of freedom. The Coulomb electrostatic interaction can be shown to arise from the exchange of virtual longitudinal and time-like photons.

longitudinal polarization **[262]** (see spin).

Lorentz boost **[138,194,284]** Lorentz transformation which does not lead to a rotation of the coordinate frame.

Lorentz factor **[116]** If an object moves with speed v with respect to a certain observer, its Lorentz factor is $\gamma = 1/\sqrt{1 - v^2/c^2}$.

Lorentz–FitzGerald contraction **[117]** Suppose that an object has a length L as measured by an observer standing still next to it. If that object moves with speed v with respect to another observer, it is seen that its length is shortened in the direction of motion by a factor $1/\gamma$, where γ is the Lorentz factor. The equation for this Lorentz–FitzGerald contraction is then $L' = L\sqrt{1 - v^2/c^2}$.

Lorentz force **[135]** Force experienced by an electrically charged particle moving in a magnetic field. The force is perpendicular to the particle velocity and to the direction of the field.

Lorentz light clock **[115]** A hypothetical instrument made of two perfectly flat, perfectly parallel, perfectly reflecting surfaces, between which a light ray bounces to and fro. Used to demonstrate the time dilatation in moving objects.

Lorentz transformation **[118]** Transformation of space-time that leaves the speed of light unchanged. A two-dimensional example: if one observer has space coordinate x and time t, and another has X and T, and if their relative speed is v, then the Lorentz transformation in space from x to X is $X = \gamma(x - vt)$, while the time part of the transformation is $T = \gamma(t - vx/c^2)$. It can be verified by straightforward algebra that this transformation leaves the 'interval' $x^2 - c^2t^2$ unchanged: $x^2 - c^2t^2 = X^2 - c^2T^2 = $ constant. The case in which the constant is zero then yields $x = \pm ct$ and $X = \pm cT$, the equation of motion for light moving with speed c in both cases.

magnetic force **[261]** In classical (Maxwell) theory, the force exerted by moving electric charges, as opposed to the Coulomb force due to stationary charges. In QED, the part of the electromagnetic force transmitted by transversely or circularly polarized virtual photons (see `Lorentz force`).

`magnetic moment` **[262]** The amount of magnetic field associated with a single particle. The quantum unit of magnetic moment is $e\hbar/2mc$, where m is the mass of the electron.

`magnetic monopoles` **[268]** Hypothetical particles predicted by some field theories. They are the magnetic equivalent of electric point charges such as the electron: a point particle consisting of a single magnetic north or south pole. These are not known to exist: so far, all magnets and magnetic particles are found to be dipoles, i.e. they have a north and a south pole that cannot be separated from each other.

`mass` **[48,204]** In classical mechanics, a measure of the inertial resistance to a force exerted on an object. Thus the force F, the acceleration a caused by it, and the mass m are related by $F = ma$, showing that railway engines are not accelerated much by a kick but footballs are. In relativity, time and space can be measured with the same unit (the second) due to the invariance of the speed of light. This has the remarkable consequence that other physical quantities, such as mass and energy, can be measured with the same unit, too. In the case of mass, this leads to $E = mc^2$ or, more properly, $E = \gamma mc^2$, where γ is the Lorentz factor of the moving object.

`mass renormalization` **[252]** The process whereby the mass of virtual particles is adjusted in such a way that the mass observed at infinity remains finite (see `bare mass`).

`matrix` **[105]** Rectangular (usually square) array of numbers which obeys certain mathematical rules and restrictions. Matrices are often sought as representations of certain groups. For example, the matrix $\begin{pmatrix} \cos\theta & \sin\theta \\ -\sin\theta & \cos\theta \end{pmatrix}$ corresponds to a two-dimensional rotation over an angle θ.

`Meissner effect` **[266]** Exclusion of magnetic field from a conductor with extremely (infinitely) high conductivity.

`meson` **[210]** Aggregate of two quarks bound together by the colour force.

`metric tensor` **[204]** A square symmetric array of 16 quantities describing the distances in a small patch of space-time. Such a distance recipe is tantamount to describing the four-dimensional structure (curvature and torsion) of space-time.

`Minkowski recipe` **[115]** Local prescription for calculating the distance between events in space-time. The distance S (called the interval) between the space-time point zero and the event (point) at time t and at spatial distance r is given by $S^2 = r^2 - c^2 t^2$.

mixed state **[94]** If a state C is a superposition of states G and L, then C is called a mixed state (see eigenstate).

mixing angle **[97,179]** Quantity describing the extent to which a given state can be seen as a superposition of two others. If the amplitude of state L is l, and that of state G is g, then a mixed state C with mixing angle θ has amplitude $c = l \sin\theta/2 + g \cos\theta/2$. Because the sine and cosine functions were originally devised to describe rotations, the rotation metaphor is used extensively.

mode **[42]** Oscillation pattern of a standing wave.

molecule **[161]** Aggregate of atoms bound together by the electromagnetic force.

momentum (conservation) **[49,105]** The product of the mass and the velocity of a particle. The total momentum of a closed system is constant (conserved). The existence and conservation of momentum is a consequence of Noether's theorem applied to the translation symmetry of space (see translational invariance).

multiplet **[102,115,175]** (of particles) Set of particles that show a certain family resemblance. The name of a multiplet is derived from the Latin word for their number: singlet for one, doublet for a multiplet of two members, triplet for three, decuplet for ten, etc. In what respect the members of a multiplet are supposed to be similar depends largely on an inspired guess. For example, the fact that the proton and the neutron are very closely similar, with the exception of their electric charge, leads one to group them together as members of an 'isospin doublet'. The similarity within a multiplet suggests devising a symmetry operation that turns one member of a multiplet into another. If such a symmetry is a gauge symmetry, this operation defines the ways in which the members of a multiplet can interact.

multiplet **[27]** (in a line spectrum) Set of spectral lines that show a certain family resemblance. In nineteenth-century spectroscopy, the resemblance was often superficial, based on the appearance of the lines; hence the indication 'S' for sharp, 'D' for diffuse, etc. Later the multiplets were seen to have a more numerical similarity, for example if they could be expressed as differences sharing the same basic terms (see Ritz's combination law).

mu-neutrino **[220]** Partner of the muon in weak interactions.

mutation **[171]** Change in the genetic makeup of an organism. Mutations have many possible causes such as attack by chemicals, energetic radiation, or copying or translation mistakes during growth (see DNA).

naked electron **[250]** Electron propagating between vertices, carrying 'bare charge and mass' (see renormalization).

neutral current **[238]** Weak interaction process without the transmission of electric charge, mediated by the Z-particle.

neutral vector boson **[236]** The electrically neutral member of the triplet of velpons that mediate the weak force, indicated by Z in this book and by Z^0 in the professional literature.

neutrino **[220]** Partner of the electron in weak interactions, forming an SU(2) doublet with it.

neutron **[28]** One of the three main constituents of ordinary matter; there are two neutrons in the nucleus of the helium atom. The neutron is a fermion with spin $\frac{1}{2}$, mass 0.9396 MeV/c^2 or 1.6727×10^{-27} kg and electric charge zero. Originally thought to be an elementary particle, it is now known to be built of three quarks in the flavour combination *udd* (see proton).

neutron star **[184]** A star that is held up against its self-gravity by the fact that its neutrons cannot be compressed into a smaller volume, thanks to the Pauli exclusion principle. A neutron star with a mass equal to that of the Sun has a radius of about 10 kilometres. A handful of neutron star matter on Earth weighs as much as Mount Everest.

node **[42]** Place on a standing wave where the amplitude is zero.

Noether's theorem **[106]** First proved by Emmy Noether: for each symmetry there is a corresponding conservation law. Example: space is invariant under translations, therefore momentum is conserved.

non-Abelian **[111]** (see symmetry group).

nucleon **[192]** Proton or neutron (isospin doublet).

nutation **[135]** Small wobble of the axis of a top that is being perturbed by a force with a component perpendicular to the rotation axis (see precession).

orbital spin **[136]** In classical mechanics, the amount of angular momentum associated with the orbital motion of a particle. In quantum mechanics, the amount of angular momentum – in units of Planck's constant $\hbar$ – associated with the angular probability distribution of a particle (see spin).

pair creation **[124]** (see antimatter).

particle **[xix, 45]** We observe in many ways that Nature is made of discrete entities. An ancient example is found in the rules of chemistry: the quantities of chemicals involved in reactions always occur in fixed ratios (e.g. one unit of oxygen plus two units of hydrogen make one unit of water vapour). When we detect matter on very small scales (such as in cloud- or bubble chambers, on scintillation screens or with scanning tunnelling microscopes) we can see this discreteness directly. The individual pieces of matter are called particles. On the small scale where quantum effects dominate, we observe families of particles; within a family, particles of a given type – such as electrons – are actually identical, something which does not occur with large things such as smoke particles. These identical particles are often called 'elementary' particles, even though there are very good reasons to believe that most small particles are made of yet smaller ones.

Pauli exclusion principle [142] No two fermions can share the same quantum numbers. Consequently, electrons in atoms must occupy higher and higher quantum states. Thus, the Pauli rule is responsible for the existence of chemistry and of white dwarfs and neutron stars (see spin and statistics).

period [32] Amount of time between successive oscillations of a wave.

periodic table [161] Mendeleyev's systematic arrangement of the elements according to their chemical affinities and atomic weights. A key feature of this table is the occurrence of the 'magic number' 8: the chemical properties of atoms with increasing weight recur every eight entries. For example, elements 2 (helium), $2 + 8 = 10$ (neon), $10 + 8 = 18$ (argon) are all 'noble gases' which have no chemical affinity; elements 3 (lithium), 11 (sodium) and 19 (potassium) are the chemically very aggressive halogens. The structure of the periodic table is a reflection of the electronic structure of the atoms, dictated by the Pauli exclusion principle. The occurrence of the 'magic number' eight can be traced to the fact that space has three dimensions, which impresses a certain regularity on the possible electron distributions. The search for similarities among particles is analogous to the building of the periodic table.

phase (shift) [32,33] The number of oscillations a wave has completed, counting from an arbitrary point, with the stipulation that whole oscillations are counted as zero. Thus, the phase can be expressed as a number beween zero and one, but usually – because of the relationship between waves and rotational motion – the phase is expressed as an angle between 0° and 360°. A change of phase is called phase shift.

photino [280] Supersymmetric partner of the photon.

photon [43,236] Particle of light, perhaps the most spectacular constituent of the everyday world. It is a boson with spin 1 and zero mass. The remarkable properties of light, in particular its wave-like behaviour (seen in interference patterns) and the fact that its speed is the same for all observers led to the theories described in this book. The photon is the velpon that derives from the local U(1) symmetry of Nature and transmits the electromagnetic force. The character of the force depends on the polarization of the exchanged photon: Coulomb electrostatic force for longitudinal and time-like polarization, magnetic force for the other two polarizations.

pion **[195]** Meson made of two quarks of the *u*- and/or *d*-flavours.

Planck length **[282]** Radius λ_P of a particle for which the Compton length equals its Schwarzschild radius. In formula, $\lambda_P = \hbar/M_P c = 2.28 \times 10^{-35}$ metres. The time needed for light to cross a Planck length is $\lambda_P/c = 7.6 \times 10^{-44}$ seconds. Phenomena on time scales shorter than this 'Planck time' cannot be described with our present-day physics (see Planck mass).

Planck mass **[282]** Mass M_P of a particle for which the Compton length equals its Schwarzschild radius. In formula, $M_P = \sqrt{\hbar c/2G} = 1.54 \times 10^{-8}$ kg, which is 10^{19} proton masses or about the mass of a bacterium. At this (comparatively speaking) enormous mass our present-day physics breaks down completely.

Planck's constant **[55,61]** Unit of angular momentum, expressed by the symbol h or $\hbar = h/2\pi = 1.05 \times 10^{-34}$ joule-seconds. The existence of $\hbar$ implies the existence of an absolute scale of smallness (see De Broglie length).

plane wave **[36]** Wave with constant amplitude and wave crests that lie in a plane.

polarization **[144]** Orientation of the oscillation vector of a wave or of the rotation axis of a spinning object. The way in which the polarization is described depends on the choice of coordinates (gauge fixing); in classical mechanics, one usually describes the polarization with respect to the direction of motion (see radiation gauge).

position **[46]** (see space).

potential energy **[49]** In classical mechanics, the amount of energy required to move a particle on which a force acts from one point to another.

precession **[129,133]** When a spinning top is pushed sideways, it responds by moving in a direction perpendicular to the rotation axis and perpendicular to the push. This is called precession, usually observed as the gyrating motion of a top placed obliquely on a surface; in that case, the push is provided by the force of gravity.

principal quantum number **[159]** In atoms, the quantum number *n* associated with the number of nodes in the radial direction of an electron distribution. The value of *n* determines the energy of this distribution. Secondary quantum numbers *l* and *m* associated with the nodes in the transverse direction (spherical harmonics) are angular momentum quantum numbers.

probability [56,67] The fraction of cases in which a particular outcome of the same experiment is seen to occur. In non-relativistic quantum mechanics, the probability is found by means of the 'wave function' ψ, which is a complex number dependent on space and time. From it, the probability distribution P in space-time can be found as $P = \psi^*\psi$, where the operation * is called 'complex conjugation'. In the Feynman formulation, each possible path is associated with a complex number called amplitude. The probability of a process is found by determining the total amplitude by summing the amplitudes of all possible paths, and then finding P the same way as one does with ψ. The total probability over all processes, including all possibilities, must be unity. This is called the 'unitarity condition'.

proton [28] One of the three main constituents of ordinary matter. It is a fermion with spin $\frac{1}{2}$, mass 0.9382 MeV/c^2 or 1.6702×10^{-27} kg and electric charge +1. Originally thought to be an elementary particle, it is now known to be built of three quarks in the flavour combination *uud*.

Pythagoras recipe [105,114] Local prescription for calculating the distance between points in space. The distance D between the space-time point zero and the point with coordinates (x, y, z) is given by $D^2 = x^2 + y^2 + z^2$.

quantized [13] A quantum particle in a stationary state behaves like a standing wave: it can only assume certain fixed configurations and is said to be quantized. This implies that small things have a shape, determined by the quantization. Some properties of particles (such as electric charge) occur only as fixed integral multiples of a basic quantity; these properties are also said to be quantized. Quantization as a verb refers to the representation of a classical force field as the sum of its constituent particles (velpons).

quantum [xx,13,64] There is a scale of absolute smallness in Nature. Small things obey a very counter-intuitive set of physical rules called 'quantum mechanics'. On this small scale, quantum effects dominate and we observe families of particles; within a family, particles of a given type – such as electrons – are actually identical, something which does not occur with large things such as smoke particles. In order to express this property of absolute smallness and its attendant physical behaviour explicitly, and to avoid confusion with large things such as soot or smoke particles, these small bits of matter or energy are often called quanta.

quantum chromodynamics [216] (QCD) Theory based on the assumption that Nature is built out of elementary triplets of particles which are connected by a local SU(3) symmetry. The particles in the triplet are called quarks; their

QCD charge is called colour; the velpon generated by the local symmetry is the gluon (see `quantum electrodynamics`).

`quantum electrodynamics` **[187]** (QED) Theory of the quantum-relativistic interaction between light and matter. It is based on the assumption that Nature is symmetric under the phase rotation operation U(1). Used as a local symmetry (gauge symmetry), U(1) generates a single velpon, the photon. This field boson couples to electrically charged particles and thereby produces the electromagnetic interaction. QED is the most exactly verified physical theory: experiments have never shown a deviation from QCD, to a precision of about 10^{-11}.

`quantum mechanics` **[66]** Theory that describes the behaviour of particles taking their wave-like properties into account; the 'mechanics of the absolutely small'.

`quantum number` **[43,102,139]** A quantum particle in a stationary state behaves like a standing wave: it can only assume certain fixed configurations. These show a characteristic pattern of nodes, which can be enumerated by means of whole numbers, just like the fundamental and the overtones of standing waves on a string. Used in an extended sense to indicate the distinguishing properties of particles such as charge, spin, etc. (see `modes`, `quantization`).

`quark` **[208]** Fundamental particle subject to the colour force, forming the level of structure below that of protons and neutrons.

`radiation gauge` **[149]** Choice of orientation for the description of the polarization of the photon, such that the direction of the spin is referred to the direction of motion (see `fixing a gauge`).

`refraction` **[90]** (see `Snell's law`).

`relativity` **[11]** Theory of motions based on global symmetry under the Lorentz group. Relativistic motions in space-time are based on the finding by Michelson and Morley (and many times since) that the speed of light is invariant: relativity is 'Lorentz-symmetric mechanics'. In a relativistic universe time and space can be measured with the same unit (the second) due to the invariance of the speed of light. Space and time are the same, as it were, so instead of treating them separately we speak of a single four-dimensional continuum called space-time. The fact that distances in space and time can be measured with the same unit has the remarkable consequence that other physical quantities, such as mass and energy, can be measured with one unit, too. In the case of mass, this leads to $E = mc^2$, or, more properly, $E = \gamma mc^2$, where γ is the Lorentz factor of the moving object (see `Lorentz transformation`).

renormalization **[250]** The process whereby the mass and the charge of interacting particles is adjusted in such a way that their values observed on large scales remain finite (see bare mass, bare charge).

representation **[104]** (of a group) Set of numbers and a mathematical operation which together obey the abstract group rules. For example, the numbers on a clock face plus the addition operation (with the stipulation that multiples of 12 are counted as zero) are a representation of the group of rotations over 30 degrees. In quantum mechanics most representations are matrix representations.

ribonucleic acid (RNA) **[171]** Close relative of DNA, made of a long strand of ribose sugar molecules linked by phosphate groups, to which side groups of various kinds can be attached. Usually, RNA is a messenger molecule or a construction tool that carries the information from DNA to the place where other biological molecules (notably proteins) are constructed.

right-handed **[145]** (see spin).

Ritz's combination law **[28]** Finding in nineteenth-century atomic spectroscopy: each atom has a set of basic frequencies (terms) from which the frequencies of spectral lines can be derived by addition or subtraction. Due to the work of Bohr this is known to be caused by quantum jumps between energy states of the electrons in the atom.

rotation **[103]** Change in orientation such that the relative positions within the rotated object remain the same and the mean position remains the same, too (otherwise we would have a translation). Example of a rotation over an angle θ about the point $(0,0)$ in two dimensions: if the original position is (x, y), the rotated position is $(x \cos\theta + y \sin\theta, y \cos\theta - x \sin\theta)$.

rotation group **[104]** Symmetry group of rotations in space (see rotation).

Rydberg equation **[101,156]** Empirical formula for the frequencies (or energies) of the spectral lines of the hydrogen atom. If E is the energy of a line, then $E = R(1/k^2 - 1/n^2)$, where k and n are whole numbers. The Rydberg constant R was calculated by Bohr, who was the first to use the quantum principles to demonstrate that the energy states of hydrogen are given by $E = R/n^2$. Thus, the Rydberg formula expresses the energy differences that correspond to photons emitted or absorbed.

S-state **[109]** Quantum state that is spherically symmetric, and therefore has no angular momentum.

scalar **[46]** A single mathematical number (see components).

scalar field **[13]** (see components).

scalar particle **[145]** Quantum representation of a scalar field: a particle represented by a single number. Thus, scalar particles have very limited degrees of freedom, and must have spin-0.

Schwarzschild radius **[281]** The radius R_S of the horizon of a spherical black hole. If the black hole mass is M, then $R_S = 2GM/c^2$. For a mass equal to that of the Sun, R_S is about three kilometres.

second quantization **[92]** Expressing a force field as a superposition of interacting field quanta (velpons) instead of as an externally imposed non-quantum force.

selectron **[280]** Supersymmetric partner of the electron.

self-energy **[250]** (see bare mass).

series **[27]** Family of spectral lines that are observed to belong together, for example by having a simple relationship between their frequencies.

singlet **[105]** Particle family consisting of one member (see multiplet).

singularity **[282]** Central space-time point of a spherical black hole.

Snell's law **[91]** When a wave moves from one medium to another in which it propagates with a different speed, the direction of the wave changes. This is refraction. Its basic cause is that a wave follows the path of least time ('Fermat's principle'). Using this, one can prove Snell's law: $\sin \alpha = n \sin \beta$, where α is the angle between the incoming wave and a line perpendicular to the interface between the two media and β is the corresponding angle of the outgoing wave. The constant n, the index of refraction, is equal to the ratio of the propagation speeds in the two media.

space **[xix]** The main property that distinguishes one particle from another. Because the Universe appears to consist of more than one particle, there must be something that distinguishes them, something that 'gets in between'. The most obvious one in everyday experience has acquired the name 'location' or 'place in space'. Descartes actually tried to define what an object is by its extent in space. There are more things than their location that distinguish particles from one another; according to some speculative modern theories, these things are the expression of a richer structure of space than mere extent.

space-like **[119]** Two points in space-time that can only be connected by a particle moving faster than light are said to have a space-like separation.

space-time **[xix,115]** Particles are distinguished from one another by many properties. Chief among these, in everyday experience, is their location in space. It is observed that the specification of the location is insufficient for a complete description (we say 'the position of particles is not always the same': the word 'always' implies the existence of the thing we call time).

Thus, we require at least one additional quantity: time. The observed fact that the speed of light is always and everywhere the same implies that space and time can be measured with the same unit, the second. Einstein showed that this means that space and time are linked in one four-dimensional continuum called space-time.

space-time diagram [9] Graphical representation of the trajectory (world line) of a particle through space-time. The non-quantum equivalent of a Feynman diagram.

special unitary group (SU) [195] Compact continuous group (Lie group) used as a local symmetry in all known quantum forces (see symmetry group).

spectral lines [26] When light from emitting atoms is sorted by wavelength (or equivalently colour or frequency) one observes that the light is seen only at or very near certain fixed wavelengths. Because of the mechanical construction of the measuring apparatus, these patches of single-coloured light have the shape of narrow strips.

spectroscope [23] Apparatus to sort and record light by wavelength (or equivalently colour or frequency), usually by means of a prism or a diffraction grating.

spectrum [25] Band of light sorted by wavelength (or equivalently colour or frequency).

speed [47] The absolute value, or magnitude, of the velocity; that is, the length of the velocity arrow (for example as expressed in metres per second) without reference to the direction thereof.

spherical harmonics [152] Pattern of standing waves on a sphere. These form a representation of the rotation group. Electron distributions in the hydrogen atom are the product of spherical harmonics and a radial function called a 'Laguerre polynomial'.

spin [19,108,137,139] The amount of angular momentum expressed in units of Planck's constant $\hbar$. When applied to orbital motion it is called 'orbital spin'; when applied to a single particle one merely uses 'spin'. The uncertainty relations limit the number of possible orientations of the spin. The orientation of the spin with respect to a fixed but otherwise arbitrary direction in space-time is called polarization. For example, a particle with spin $\frac{1}{2}$ can have spin 'up' or 'down'. Usually, one takes the direction of motion of the particle as the direction of reference. When the spin corresponds to a right-handed screw moving in the direction of motion, we speak of right-handed spin; the opposite is left-handed. The handedness of the spin (R or

L) is called helicity. Alternatively, we can add or subtract the helicities; this gives us two transverse polarizations. The direction perpendicular to these in three-space corresponds to a longitudinal polarization (i.e. in the direction of motion). In four-dimensional space-time, the spin can be oriented in the direction of the time axis; this is called time-like polarization.

`spin and statistics` **[142]** (law of) Collective name for Bose–Einstein statistics and Fermi–Dirac statistics. Due to the rotational symmetry of space and to the quantum behaviour of Nature, particles come in two types: those that have integral spin (bosons) and those with half-integral spin (fermions). It can be shown that this implies a profound difference in the way indistinguishable particles interact. No two fermions can share the same quantum numbers, whereas bosons are more likely to be found with equal quantum numbers. Given one fermion in a quantum state, the probability that another identical one will find a place in that state is zero. Thus, fermions are 'intolerant'. Given N identical bosons in a quantum state, the probability that yet another one will find a place in that state is multiplied by a factor $N + 1$. Thus, bosons are 'gregarious' particles. The law of spin and statistics is seen most vividly in our everyday distinction between matter (fermions) and forces (bosons).

`spinor` **[145]** In its simplest form, a four-component vector built up out of two two-component vectors. The four-vector can be used as a relativistic (Lorentz) vector, whereas a two-component vector cannot (2 is not a power of 4). For example, an electron and its antiparticle, the positron, can be represented by a spinor.

`spontaneously broken symmetry` **[227]** A solution of an equation of motion that is manifestly symmetric need not itself show that symmetry. For example, the laws of planetary motion are spherically symmetric and yet planets orbit on ellipses in a plane. This is one example of spontaneous symmetry breaking. On a quantum level, the self-interaction of particles can lead to spontaneous symmetry breaking.

`squark` **[280]** Supersymmetric partner of the quark.

`standard model` **[273]** The assembly of quantum field theories that describe the behaviour of all known particles: the electroweak theory based on spontaneously broken local U(1)⊗SU(2) symmetry, and QCD based on local SU(3). Up to 1994 all known experiments can be accommodated within this model, but gravity is absent and other nagging deficiencies remain.

`standing wave` **[41]** An oscillation pattern that occurs when a wave is constrained to oscillate with fixed boundary conditions.

`state (vector)` **[93]** List of properties of a quantum, usually indicated by an enumeration (vector) of the relevant quantum numbers. For example, an electron in a hydrogen atom can be in the $n = 3$, $l = 2$, $m = 0$ state. The ground state of a given interaction is the state with lowest energy; states with higher energies are said to be excited states.

strong force **[192,210]** Force between mesons and baryons, due to deformation of their quark clouds when they come closely together. It is a residual force, arising because of small local changes in the balance of the colour forces among quarks. The strong force in ordinary nuclei can be described by the Yukawa model, i.e. the exchange of lumps of quarks (pions) between bigger lumps of quarks (baryons). The strong force holds atomic nuclei together against the electromagnetic repulsion among the protons.

subgroup **[109]** Subset of a symmetry group, which itself obeys the group rules. For example, the rotations over right angles form a subgroup of the rotations over 30 degrees.

supergravity **[281,286]** Unverified theory of the interaction between particles, including gravity, based on the use of the super-Poincaré group as a local symmetry.

super-Poincaré group **[279]** Generalization of the Lorentz group. The super-Poincaré group not only includes, like the Lorentz, rotations in space-time and a mapping that leaves the speed of light invariant, but it also encompasses translations. This makes the super-Poincaré group useful for the description of motion which, after all, is a form of repeated translation.

superposition **[37]** Linear addition of the amplitude of waves. If the amplitude is a single number (such as in water waves), these numbers are simply added. If the amplitude is more complicated, the appropriate addition laws must be used, for example with the complex numbers of quantum amplitudes (these can also be treated as concatenating amplitude vectors tip-to-tail).

superstring **[287]** Hypothetical object that describes particles as quantum states of a field in a dimension of the vacuum other than the usual space-time four.

supersymmetry **[279]** Unverified theory that allows bosons to turn into fermions, and vice versa, by means of a postulated supersymmetry. It has the consequence that all particles have a supersymmetric partner. The bosonic partners of fermions are indicated by placing an s in front of the name (e.g. squark, selectron); the fermionic partners of bosons are indicated by adding the ending -ino to the word (photino, gluino). There is no evidence at all that such particles exist.

symmetry, symmetric **[xx,18,101,104]** If something remains unchanged under a certain operation, it is said to be symmetric under that operation. For example, a sphere is symmetric under all space rotations around a fixed point; a cube is symmetric under right-angle rotations about its centre; space is symmetric under translations, i.e. every point in space is as good as any other. The fact that a symmetry necessarily leaves something unchanged finds a deep expression in conservation laws (see Noether's theorem).

symmetry group **[101,103]** A symmetry group is a set of objects, called elements, plus some prescription for relating these elements to each other. Group elements are related as follows: (1) With every pair of elements is associated a third one. The association is known as group multiplication, or multiplication for short, indicated by $\otimes$. From two elements a and b we can form $a \otimes b = c$. The elements a, b and c need not be different, nor do we require that $a \otimes b = b \otimes a$. (2) The multiplication is associative; that is to say, if we chain three elements together by multiplication, it does not matter which of the two multiplications we perform first. In symbols, this means that $(a \otimes b) \otimes c = a \otimes (b \otimes c)$. (3) There is a unique element of the group, called the unit element (usually indicated by e or I), such that for each element a one has $a \otimes e = e \otimes a = a$. (4) Each element a has an inverse (usually called a^{-1}) such that $a \otimes a^{-1} = e$. A set of numbers and a mathematical operation which together obey the abstract group rules is called a representation of that group, for example the integer numbers $(1,2,3,\cdots$, their negatives and zero) form a group under ordinary arithmetic addition. If the smallest object that represents a group is an $N \times N$ matrix, then N is called the dimension of the group (such as three in SU(3)). A group for which $a \otimes b = b \otimes a$ is commutative or Abelian. An object that does not change when operated on by any group member is symmetric or invariant under that group. When a symmetry group is used as a local quantum symmetry, it is usually called a gauge group.

tensor field **[13]** (see components).

tensor particle **[145]** Particle with whole-numbered spin higher than one: $2, 3, 4, \cdots$.

terms **[28]** Set of basic frequencies from which the frequencies of spectral lines can be derived by addition or subtraction (see Ritz's combination law).

time **[xix,46]** It is observed that the arrangement of objects in space is not enough to describe them. We require at least one additional quantity, historically called time. The observed fact that the speed of light is always and everywhere the same implies that space and time can be measured with the same unit, the second. Einstein showed that this means that space and time are linked in one four-dimensional continuum called space-time.

`time dilatation` **[117]** The change of the rate at which a clock ticks as seen by an observer moving with speed v with respect to that clock. In formula: if t is the time between ticks observed when standing still next to the clock, then T is the time between ticks observed when reading the moving clock, where $T = \gamma t$ and γ is the Lorentz factor (see `Lorentz–FitzGerald contraction`).

`time-like` **[119]** Two points in space-time that can be connected by a particle moving more slowly than light are said to have a time-like separation.

`time-like photon` **[262]** Photon polarized in such a way that its spatial components are zero. Relativity shows that, because a photon has no rest mass, a real (free) photon cannot have time-like polarization. However, the uncertainty relation allows a virtual photon to have a time-like degree of freedom. The Coulomb electrostatic interaction is due to the exchange of virtual longitudinal and time-like photons.

`translation` **[103]** A change of position in space, in such a way that the orientation of the translated object does not change. Any motion in space can be composed of a translation followed by a rotation, or vice versa.

`translation group` **[104]** Symmetry group of translations in space. For example, a sheet of square graph paper shows translational but not rotational symmetry. Due to the fabrication process, most cotton print patterns are translationally symmetric.

`translational invariance` **[105]** Something that remains the same when translated is invariant under the translation group. All things are observed to be translationally invariant: when riding in a train you don't change into a pumpkin or a rabbit (complain to the conductor if you do). The conserved quantity corresponding to this invariance is the momentum.

`transverse spin` **[145]** (see `spin`).

`uncertainty relations` **[61]** There are pairs of physical quantities, called complementary or conjugate pairs, with the property that the precise knowledge of one of them precludes all knowledge of the other. The uncertainty relations are mathematical expressions of this complementarity. For example, the position x and the momentum p are a conjugate pair; the uncertainty relation says that the precision Δx and Δp with which x and p can be determined are related by $\Delta x \times \Delta p > \hbar$, where $>$ stands for 'must be greater than'.

`unification` **[269]** Demonstration that several seemingly different forces are in fact different aspects of a single force. The first example of unification is due to Newton, who showed that the force that makes an apple fall to the ground is the same as the force that keeps the Earth and the Moon in orbit around each other. Maxwell and Faraday unified electricity and magnetism.

unit element **[104]** There is a unique element in each symmetry group, called the unit element (usually indicated by e or I), such that for each element a one has $a \otimes e = e \otimes a = a$ (see symmetry group).

up quark **[210]** (see flavour).

up spin **[146]** (see spin).

vacuum **[xx]** Professional slang for what non-physicists call 'space': the lowest-energy state of space-time with all its resident fields. The term is meant to express that space-time with its relativistic fields is quite a lively place. For example, the vacuum can be asymmetric, in the sense that a vacuum with particles in it is more likely than a vacuum that is empty. The vacuum shows polarization effects which help in determining the way in which particles propagate, and so contribute to their mass and charge.

vacuum polarization **[255]** Due to the interplay between the uncertainty relations and relativity the number of particles in an interaction is not a constant: pair creation and annihilation are possible everywhere. Thus, one cannot say that a region of space-time is 'empty'. Spontaneously appearing and annihilating particle pairs give an active role to the vacuum. The effect this has on the propagation of particles is called vacuum polarization (see antiparticle).

Van der Waals force **[184,217]** Subtle and weak force between atoms, molecules and bigger objects, due to deformation of their electron clouds when they come closely together. It is a residual force, arising because of small local changes in the balance of the electrical forces between electrons and atomic nuclei.

vector **[46]** A row of numbers obeying certain mathematical rules and regularities (see components).

vector boson **[145]** (see vector particle).

vector field **[13]** (see components).

vector particle **[145]** Quantum representation of a vector field: a particle represented by a row of numbers. In relativistic theories the vector must have four components because of the four dimensions of space-time. Accordingly, vector particles have spin-1 (vector boson). This gives them greater freedom than scalar particles, and the force they transmit as a velpon is more complicated than a scalar force. Vector bosons can be derived from complicated transformation rules in symmetry groups.

`vector potential` **[190]** Field density of the electromagnetic field (see `field density`).

`velocity` **[32,46]** The rate at which the position changes with time: the *first time derivative* of the position. Determined just the way the police find out if you're speeding: observe a position, do the same an instant later, subtract the positions (by means of vector subtraction) and divide the difference by the time difference between the two positions. Thus, the velocity is a vector with the same dimension as the position.

`velpon` **[183]** Generic name for the various vector bosons which transmit quantum forces, after a Dutch type of glue purported to stick to everything (see `vector boson, gluon, photon`).

`vertex` **[15]** The place where, in a Feynman diagram, fermions and bosons share a point in space-time.

`virtual quantum` **[18]** Particle connecting two Feynman vertices. Thus, it is not seen outside but occurs only within an interaction. But because the observed interaction is the sum of all its Feynman alternatives, virtual particles do have an influence on the result seen at large.

`wavelength` **[32]** Distance between successive wave crests, usually given the symbol λ.

`wave packet` **[60]** Superposition of waves with various wavelengths in such a way that the resulting amplitude is appreciably different from zero in a small region of space only.

`wave vector` **[63]** Vector pointing in the direction of propagation of a wave; its length is inversely proportional to the wavelength. In relativity, the wave vector has a fourth vector component, proportional to the frequency.

`weak charge` **[235]** A measure of the strength of the 'weak' nuclear coupling between particles; a consequence of Noether's theorem applied to the internal SU(2) symmetry of particles (see `charge`).

`weak force` **[222]** Originally identified as the force between electrons and neutrinos, and responsible for nuclear beta decay processes. Later seen to act between quarks, too (see `electroweak force`).

`white dwarf` **[184]** A star that is held up against its self-gravity by the fact that its electrons cannot be compressed into a smaller volume, thanks to the Pauli exclusion principle. A white dwarf with a mass equal to that of the Sun has a radius of about 10 000 kilometres.

Yang-Mills field **[195]** Velpon field that arises when the isospin symmetry SU(2) is used as a local symmetry. Because of the dimension two, there are $2^2 - 1 = 3$ velpons in this field. Yang and Mills meant to identify these with the three pions, but they saw that this is impossible because pions are not massless and do not have spin-1, as is required for velpons. But all modern local-symmetry field theories are of the Yang–Mills type.

Yukawa's law **[264]** Expression for the effective force due to the exchange of a massive scalar particle. If r is the distance between particles, then the Yukawa force can be derived from the potential $Y = \frac{1}{r}e^{-r/\lambda_C}$. The number λ_C is the Compton length of the exchanged particle. The factor $1/r$ is the same as in the Coulomb potential; the exponential factor e^{-r/λ_C} is due to the fact that energy must be 'borrowed' to create mass, which limits the range of a massive virtual particle.

Index